OPTIMAL CONTROL OF CREDIT RISK

Advances in Computational Management Science

VOLUME 3

The titles published in this series are listed at the end of this volume.

Optimal Control of Credit Risk

by

Didier Cossin
Université de Lausanne, Switzerland

and

Felipe M. Aparicio
Universidad Carlos III de Madrid, Spain

Kluwer Academic Publishers
Boston/Dordrecht/London

Distributors for North, Central and South America:
Kluwer Academic Publishers
101 Philip Drive
Assinippi Park
Norwell, Massachusetts 02061 USA
Telephone (781) 871-6600
Fax (781) 871-6528
E-Mail <kluwer@wkap.com>

Distributors for all other countries:
Kluwer Academic Publishers Group
Distribution Centre
Post Office Box 322
3300 AH Dordrecht, THE NETHERLANDS
Telephone 31 78 6392 392
Fax 31 78 6546 474
E-Mail <orderdept@wkap.nl>

Electronic Services <http://www.wkap.nl>

Library of Congress Cataloging-in-Publication Data

A C.I.P. Catalogue record for this book is available from the Library of Congress.

Printed on acid-free paper.

Printed in the United States of America

Contents

Acknowledgments

We thank Robert C. Merton, Carliss Y. Baldwin, Jamil Baz, Franois Degeorge, Jean Dermine, Athanasios Episcopos, Kenneth Froot, Van Son Lai, Scott Mason, Ernst-Ludwig von Thadden, conference or seminar participants at Harvard Business School, INSEAD, ESSEC, University of Geneva, George Washington University, INCAE, FMA, and EFA for discussions and comments; Linda Loehr for copy editing. Financial support was provided by the Division of Research, Harvard Business School and HEC, University of Lausanne. We remain entirely responsible for possible errors.

Chapter 1

INTRODUCTION

Guarantees, whether implicit or explicit, are a widespread component of financial contracts and provide the conceptual framework for credit risk analysis. Foremost among explicit private guarantees are guarantees of the debt obligations of subsidiaries from parent corporations; letter-of-credit guarantees provide by commercial banks; swap guarantees; mortgage guarantees; and insurance contracts of all sorts. Public guarantees are ubiquitous. Government guarantees of loans made to private corporations have made headlines on each side of the Atlantic. Indeed, government-issued small business guarantees as well as export-oriented and industry-targeted guarantees represent current government practices for financing economic activity. Even more important may be the role of guarantees of deposits through the Federal Deposit Insurance Corporation (FDIC), and less extensive, but similar, guarantees of pension benefits, student loans, residential mortgages, etc.

But implicit guarantees are also widely used. As a matter of fact. any risky loan can be considered a combination of a riskless contract and a guarantee, as shown by Merton and Bodie (1992).

The use of guarantees is likely to become even more widespread in the future. The enormous amount of recent financial investment products highlights the importance of contract fault or credit sensitivity, as it appears for example in swap trading.

In general, any form of default or signals the existence of an implicit guarantee. Starting with Merton (1974) up to a flow of recent papers with different approaches, the academic literature has considered credit risk mostly as a pricing issue. The challenge has been to find the right model to price credit risk and many have added to the theory from this side, either by following and extending Merton's approach (Shimko and alii (1993), Longstaff and Schwartz (1995), Anderson and Sundaresan (1996), Leland and Toft (1996), Mella-Barral and Perraudin (1997) amongst others) or by choosing an alternative path (Duffie and Singleton (1996), Duffie and Huang (1996), Jarrow and Turnbull (1995), Jarrow, Lando and Turnbull (1997) amongst others). See Cossin and Pirotte (2000) for a survey of credit risk models.

Credit risk pricing, following Merton's terminology, corresponds to guarantee pricing as any long position in a credit risky contract can be considered identical to a portfolio of a long position in a riskless contract and a short position in a guarantee (Merton and Bodie, 1992). As such, the pricing research on credit risk is valid, with slight modification to fit the institutional context, for pricing the guarantees that pervade financial contracts, such as letter-of-credit guarantees, mortgage guarantees, swap guarantees, public guarantees (such as for deposit insurance, pension benefits, etc.), and all the implicit embedded guarantees that appear in risky contract (see Hirtle, 1987, and the Appendix of this monograph for a survey of guarantees). Many applications using traditional financial engineering on the Merton approach to pricing credit risk and guarantees have been developed for each of this situations (e.g., Jones and Mason, 1980, on classical debt, Cooper and Mello on swaps, 1991, Hsieh and alii, 1994, on PBGC insurance premiums, Marcus and Shaked, 1984, and Crouhy and Galai, 1991, on deposit insurance) .

Banks and other financial intermediaries, on another hand, have for a long time mitigated their credit risk exposures (or other short positions in guarantees, to use the Merton terminology) by using collaterals rather than pricing them. This is particularly evident in the OTC markets, e.g. the swap market, where counterparties of different quality main not obtain different rates but will often be asked for a collateral. It is well known that collaterization substitutes (at least in part, see Cossin and Pirotte, (1997) on swap pricing and Cossin and Hricko (2000) for theoretical results in the field) for credit risk pricing. In our knowledge, very little academic research bears on what an optimal policy of collaterization should be (some research analyzes the role of collaterals in game-theoretical models of debt renegotiation -see Bester, 1994).

Given that credit risk and guarantees are pervasive financial instruments, their optimal control can be an important competitive advantage for companies, while in the case of public finance, it may result in significant savings for taxpayers. In many cases, the option-based valuation approach to guarantees could not be directly implemented or sufficient to manage the risks involved (as in deposit insurance). Therefore, the guarantor has somehow to devise policies to keep under control the value of the assets underlying the guarantee. Knowing that the actions allowed on the part of the guarantor are essentially of two types, namely a) to monitor these assets, possibly with cost, and b) to control them with cost (for example, seize them under some circumstances or restrict their use, or just ask for additional collateral), the programs implementing these policies should provide the guarantor with the optimal timing and strength of its controls. Current general practice relies on standard rules of thumb to monitor the assets, tracking them through annual or quarterly audits(or even daily,as for example in margin accounts). Alternatively, seizure or control policies of assets securing guarantees have been based on static capital ratios, with some discretionary power often used by the guarantor to skirt these rules (see for example the 1978 guidelines set by the Controller of the Currency, the Federal Reserve and the FDIC, so-called CAMEL ratings). But none of these procedures are enough to ensure profit maximization. The same can be said about premiums based on the historic value of the assets (even those calculated using refined theories such as contingent claim analysis). Indeed, fair premiums are valid in so far as neither the guaranteed party nor the guarantor can influence the dynamic path of the assets or liabilities backing the guarantee without the immediate knowledge of the counterparty. In a realistic agency setting, however, a moral hazard problem arises since the guaranteed party has strong incentives to alter the stochastic process of the asset's prices, either by hiding its true value or by increasing its variance so as to inflate the value of the corresponding put option. Thus unless the guarantor can monitor the assets continuously, risk-based premiums will be insufficient to ensure that the requested premium is in line with the guarantor's cost. Moreover, for regulatory or marketing reasons, it is not that easy to implement risk-based premiums. There are many cases in which only a global approximation of risk, or a discrete approximation such as the risk categories in insurance, is used in calculating the premium. But in addition, the available data history could be affected by different sources of errors or inaccuracies.

It follows from the previous considerations that the efficiency of offering guarantees depends on optimal monitoring and optimal control. Here, we propose a set of optimization programs that solve the control problem by first characterizing the minimum cost of guarantees and then the optimal behavior leading to that cost. Even though they represent a first attempt at modeling what should impact the decision to ask for more or less collateral during the life time of a contract or guarantee, our programs will remain valid for the control of any financial contract at risk of default (swaps, CMOs, etc.), since as we said before, any risky loan can be decomposed in a safe contract and a guarantee. We treat our optimal control problem following the methodology of *impulse control* proposed by Bensoussan and Lions (1973). Impulse control techniques are not frequently analyzed in the available financial literature and should prove useful to banks and financial intermediaries that want to optimize their collaterization policy (notably when it bears on large amounts, for example for AAA set-ups). In this respect, our work proposes a new framework that should be thought of as a basis to be refined for practical use. The formulation of the original singular stochastic control problem lead us to a set of quasi-variational inequalities (as developed in Bensoussan and Lions, 1982; Glowinski, Lions and Trémolières, 1981; and Hlavácek and Alii, 1988) which we try to solve numerically in a simple set-up. We finally discuss the type of results provided by our analysis.

The monograph is organized as follows. In Chapter 1, we start by reviewing the main avenues of literature related to our problem. Chapter 2 provides a brief overview of the main optimal control principles which we use later, while Chapter 3 presents the models and their setting. Here we define the relevant variables, the model parameters, and the solution approaches to be considered. In the remaining chapters, we propose two sets of programs. One set of programs will apply in cases where the information on the assets' value is readily available (*full observation* case), while the other applies when costly audits are needed in order to assess this value (*partial observation* case). Chapter 4 deals with the full observation case, first as an impulse control problem and then as an optimal stopping problem. Chapter 5 proceeds similarly for the partial observation case. In either case, the modeling stage lead us to a set of *quasi-variational inequalities* which we attempt to solve numerically in the simpler case of full observations. This is done in Chapter 6. Finally, a simulation analysis is carried out in Chapter 7, in which we study the influence on the control process of changes in the different model parameters. This precedes a discussion on possible extensions in Chapter 8 and some concluding remarks in Section 9. An Appendix provides for a survey of different situations involving financial guarantees and the need for optimal control.

Chapter 2

LITERATURE REVIEW

Various strands of financial and mathematical literature relate to the research on the control of guarantees. On guarantees, the main corpus of academic research focuses on valuation, with special attention given to deposit insurance valuation . Research on the control of guarantees is more limited, often focused on optimal seizure rules for regulators, with the exception of a paper studying general management of guarantees (Merton and Bodie, 1992) and of another one studying the impact of collateral on the pricing of guarantees (Cossin and Hricko, 2000).

The study here does not rely exclusively on the guarantee literature; because of the technology involved, it also uses the literature of portfolio selection with transactions costs as well as the literature of machine quality control. The following section describes these different strands of literature as they relate to this study.

1. GUARANTEE VALUATION

The financial, political, and institutional stakes of accurately valuing guarantees have long been recognized. Indeed, an exact theory of guarantee valuation is now well developed. Merton (1977) initiated the use of contingent claim analysis to value guarantees, with an application to deposit insurance. He noted that guarantees are isomorphic to put options, with particular maturities and exercise prices, as well as underlying processes. For example, Merton presented a model in which the price of deposit insurance (as a specific example of guarantee) is equal to the value of a put option written on a bank's assets with a strike price equal to the amount of insured deposits. He assumes in this model that the guarantor examines a bank at specified time intervals, hence obtaining from these time intervals the maturity of the claim. In another study, Merton (1978) derives what would be a fair one-time payment by banks (i.e., a chartering fee) for deposit insurance, given that there is a cost in auditing banks at random time intervals. Merton's two original papers (1977 and 1978) influenced many subsequent studies, of which a few are surveyed here (see in particular Jones and Mason (1980); Sosin (1980); Langetieg et alii (1982); Bulow (1981 and 1983); Harrison and Sharpe (1982); Baldwin, Lessard, and Mason (1983); Bulow and Scholes (1983); Marcus and Shaked (1984); Flannery and James (1984); Ronn and Verma (1986); Pennacchi (1987a and b); Marcus (1987); Thomson (1987); Selby, Franks, and Karki (1988); Cummins (1988); Bodie (1991); Crouhy and Galai (1991). Among early uses of option pricing in insurance valuation, note also Sharpe (1976); Mayers and Smith (1977); Treynor (1977).).

Specific analyses to a wide range of guarantees followed. Among these, Jones and Mason (1980) use numerical solutions (obtained by the method of Markov chains) to value a range of guarantees: full guarantee of non-callable coupon debt; partially guaranteed issue of non-callable coupon debt; junior and senior non-callable debt with guarantees; callable coupon debt. Sosin (1980) extended this approach by recognizing potential wealth transfers to the original stockholders from the guarantee. Regarding as irrational the tendency for equity holders to redistribute wealth away from themselves, he assumes a no-loss, no-gain condition for existing bondholders (for his analysis, Sosin considers a politically driven, non-profitable project, that decreases the value of the firm (p.1212). He then assumes that protective covenants of senior debt holders prevent redistributions away from them in order to allocate the value of the claims on the firm value between new equity holders, old equity holders, subordinated debt holders and senior debt holders (p.1213). Because "it is irrational for equity holders to knowingly redistribute wealth away from themselves" (p.1213), the firm recapitalizes at the time of investment.)

Baldwin, Lessard and Mason (1983) warn against the "budgetary time bombs" constituting uncontrolled guarantees offered by the government (in that case, the Canadian Government) and urge governments to account for the fair value of guarantees as measured through contingent claim analysis. Such accounting is now mandated by law (1990) in the United States through the Office of Management and Budget and in Canada, but not, in our knowledge, in Europe.

Selby, Franks and Karki (1988) value loan guarantees and the wealth transfers that may arise as a result of the changes in capital structure introduced by the new guaranteed loan. They assume a no-loss, no-gain condition for shareholders more fitting to the United Kingdom context than Sosin's no-loss, no-gain condition for bondholders. This condition arises from different bankruptcy rules. In the United Kingdom, when a firm is in financial distress, a trustee (known as the "receiver") can be appointed by the lenders to manage the firm. As a result, wealth transfers against shareholders may be difficult to reverse. In the United States, however, such a firm can seek protection from the courts under Chapter 11 of the 1978 Bankruptcy Act; only rarely is existing management removed. In the specific case of ICL, the UK government did not require renegotiation of any of the outstanding debt (while lenders to Chrysler redeemed a third of their debt at a price substantially below par). The authors also use compound options to deal with an outstanding hierarchy of loans with more than one maturity. They then apply their valuation theories to the valuation of the loan guarantee provided in 1981 by the United Kingdom government to International Computers Limited (ICL). Cummins (1988) develops risk-based premium formulas for insurance guarantee funds, both for ongoing insurance with or without jumps in assets behavior and in policy cohorts (where liabilities eventually run off to zero as claims are paid). He shows how the use of risk-based premiums can relieve the adverse incentives of firms arising from the use of non-risk-based or "flat" premiums.

Cooper and Mello (1991) develop a partial equilibrium model for swap default that can provide banks with a measure of their swap transactions net of default risk, and regulators with a consistent way of measuring the potential default risk to effectively control banks. They characterize the transfers arising among shareholders, debtholders, and swap counterparties and obtain closed-form solutions for the value of the default risk in the swap. They do not, however, consider collaterization of swaps. Bodie (1991) examines the guarantee provided by the sponsor of a defined benefit pension plan by referring to its equivalent put option and discusses the optimal hedging strategies that the plan sponsor should follow.

Cossin and Hricko (2000) address the issue of pricing credit risk with the specific guarantee of a collateral in a structural form framework. Notice that the issue of pricing an instrument that is collateralized with another risky instrument is not trivial and becomes complex when marking-to-market or margin calls are considered. Margrabe(1978) has mentioned the analogy between an exchange option and a margin account and provides the pricing for a very simple framework with no marking-to-market. Stulz and Johnson (1985) have priced secured debt using contingent claim analysis and study the use of collaterization in a corporate finance framework, analyzing the impact of collaterization on the value of the firm. The rest of the economic literature has addressed the rationale behind the use of collateral in debt contracts and is an extension of the questions arising in the theory of debt (see Benjamin, 1978; Plaut, 1985; Bester, 1994) but has not been concerned with pricing the credit risk with collateral or with evaluating the impact of haircut levels on the credit risk value.

2. DEPOSIT INSURANCE VALUATION

The issue of deposit insurance pricing dominates the literature on guarantee valuation. As already noted, the original Merton papers (1977 and 1978) focus on that particular application.

Marcus and Shaked (1984) use Merton's (1977) model to estimate from bank-stock market data the fair value of deposit insurance for a sample of large commercial banks. They find that the vast majority of larger banks (publicly traded as they use market data) are overcharged for deposit insurance. Marcus and Shaked assume, however, that regulators could control the banks (i.e., implement new capital requirements) at no cost as they valued the put option (the security equivalent of the deposit insurance guarantee) from one audit to another. In other words, they assume that just after the next audit, the value of the FDIC's liability is 0 because, at that time, regulators can implement full control.

Pennacchi (1987b) generalizes Merton's model and considers alternative policy assumptions concerning an insuring agency's pricing of insurance (i.e., fixed rates over different risk classes versus risk-sensitive rates) and methods for handling bank closings by the agency (direct payment to depositors versus arrangement of a merger). Pennacchi also describes incentives for risk-taking by the banks under the different policies.

Ronn and Verma (1986) present an empirical valuation methodology also based on Merton (1977), using only market data on equity to determine asset value and volatility (by inverting the valuation of equity as a call option on assets), rather than any data provided by bank management or by FDIC audits. Kane (1985, 1986) explores the consequences of valuing deposit insurance incorrectly and the policy implications of using contingent claim analysis. He also discusses the constraints faced by FDIC that may prevent it from closing insolvent banks, allowing "zombie banks" to persist.

Flannery and James (1984) use stock market data to obtain effective maturities for different bank liabilities such as demand deposits, regular savings accounts, small denomination time deposits. Use of their methodology may help determine values of puts on these liabilities. Thomson (1987) uses market information to value deposit guarantees. He decomposes the FDIC guarantee into the guarantee of insured deposits, a conditional guarantee of the bank's uninsured deposits, and a guarantee of stockholders' residual on the future earnings of the bank (since the FDIC may fail to close out the stockholders' position in an insolvent bank).

Crouhy and Galai (1991) use contingent claim analysis to develop a model of a financial intermediary and obtain the valuation of equity and demand deposits for the intermediary. They also obtain valuation for fair deposit insurance premium and further analyze the impact of different types of regulation for control of the assets (e.g. reserve requirements, capital requirements, interest rate ceilings, etc.) on the banking industry.

3. CONTROL OF GUARANTEES

The formal literature is more scarce on the control of ongoing guarantees. A special branch of the literature focuses on the optimal seizure of banks by regulators.

Campbell and Glenn (1984) focus on the determinants of an optimal price for deposit insurance and optimal policy for closing insured institutions, analyzing the practicality of a private deposit insurance system. In no way, however, do they provide the means of obtaining this optimal closure policy. Their study focuses on measurement problems and the impact of either early or late closure, and the control problems arising from the closure rule itself.

Pennacchi (1987a) was the first to formally analyze the link between pricing of deposit insurance and optimal control. He studies the impact of regulatory control in insuring deposits over the valuation of the insuring's agency liability. He examines two cases of full control (following any audit, regulators can compel a capital deficient bank to add more capital) and of no control (regulatorscan close a bank only when it has negative net worth and cannot influence its capital structure before that). The two cases can also be understood as being limited-term insurance for the first case and unlimited term for the second. Audits occur in his model at Poisson distributed random time intervals (as in Merton 1978) and in a non-endogenous manner. He shows that for a large sample of banks, the FDIC should be seen as overcharging the banks if actual control is full control and as undercharging them if actual control is closer to no control.

Acharya and Dreyfus (1989) derive optimal regulatory policies for closing or seizure of a bank by FDIC as functions of rate of flow of bank deposits, interest rate on deposits, the economy's risk-free interest rate, and the regulator's audit/administration costs. Baldwin (1991) analyzes the phenomenon of "asset stripping" and its impact on the cost of deposit insurance when there is information asymmetry between management and deposit insurance agencies. She discusses regulatory control of banks and the source of strength doctrine, all elements that would have to be considered in a direct application of our model to deposit insurance. Additionally, she shows how new equity infusions or secured interests in specific assets (collaterals) ought to be the modern regulatory response to bank diversification and the rise of secondary markets. Both solutions can be modeled following our program, resulting in optimal control.

Gennotte and Pyle (1991) provide an analysis of effects of deposit guarantees on banks' loan portfolios when there is imperfect regulatory control of the banks' assets from the guarantor. They show how deposit guarantees lead banks to engage in inefficient investment and how tightening control can actually lead to an increase in per-unit asset risk (and thus possibly to an increase in the probability of default).

Merton and Bodie (1992) treat the management of guarantees as a general issue, clarifying in the process the context in which the guarantee control problem arises. Further, their discussion specifies when and how the optimal behavior derived in our paper can be implemented. They show why guarantees, whether they are explicit or not, need to be managed. The institutional structure in which guarantees originate (government versus bank, for example) certainly affects the management issue but does not preclude the issue from arising in any way. Stressing the different dimensions of guarantee management, Merton and Bodie show that profitability of guarantee management arises by combining the right mix of adequate premiums with control of operating costs as well as the frequency and severity of shortfall losses. They categorize the three tools of guarantee management as (1) restriction of types of assets, (2) the monitoring and seizure of assets, and (3) risk-based premiums. Any form of guarantee management is a mix of these three tools. Depending on regulation, taxes, the institutional structure, or the structure of information, some of these tools will be preferred. In some cases, information is so cheap to obtain, for example, that monitoring almost exclusively is an excellent solution. (See the margin account example.) In others, risk-based premiums will not be allowed because of regulation, so the other two forms of management must dominate. (See the deposit insurance example.) Valuation theory addresses only the premium calculation part. Still needed is a formal theory of the monitoring and control timing. When there are costs involved in monitoring the guaranteed assets, continuous monitoring is not feasible at finite cost. A trade-off occurs between the cost of frequent monitoring and the possibility of the collateral assets falling substantially between audits, thus jeopardizing the guarantor's position.

Our perspective here is to build programs determining (1) when a guarantor should audit a company for which it provides guarantees, and

(2) when it should control the assets' level to minimize its costs over time. This approach differs from previous studies (such as Pennacchi,1987a; Acharya and Dreyfus,1989; Gennotte and Pyle, 1991) because it presents a general rather than specific framework and because it proposes a mathematical technology still rarely used in finance (and never used for that particular problem).

Note that unexpected changes in assets level or variance can be linked to problems arising from asymmetric information between the guarantor and the guaranteed party. Moral hazard, adverse selection, or agency problems can all arise in different situations (e.g., moral hazard is a problem is the company buys a put from the guarantor and then can increase the value of the put by increasing the variance of the assets). Our programs partly help to overcome these problems. This monograph allows both for situations where there is full information (no cost to audit) and partial observation(costly audits). It allows for multiple controls applied at various times during the guarantee's lifetime (impulse control) as well as for a onetime application of control, such as seizure of a bank by a regulator (stopping time). The programs presented here can be implemented directly, although applications to real-world situations will increase both conceptual and mathematical complexities already met here. No unusual data are necessary. Modern computers are required to handle the program through numerical analysis. Guarantors who today apply rules of thumb for timing their (often very costly) audits may gain by similar programs to optimize their behavior.

4. OTHER APPLICATIONS

The problem of guarantee control as analyzed here is similar to both the problem of portfolio selection and the problem of machine-quality control, as explained in the following sections. When there are fixed costs to transacting or to controlling (depending on the case), the three problems are known in mathematics as "impulse control" problems. We show the link between the different approaches to "impulse control" and the origin of the "quasi-variational inequalities" technology, as developed in Bensoussan and Lions (1982) and used in the analysis to follow.

The problem of optimal guarantee control is homologous in many ways to the problem of portfolio selection when there are fixed transactions costs. The portfolio selection program of an investor can be written as

$MaxE\left\{\int e^{-\delta t}.u(c(t)).dt\right\}$ with t the time between 0 and maturity T, $u(.)$ the utility function of consumption $c(t)$ and δ the discount rate.

This program can be easily compared to the objective function of the complete observation program developed later. When considering only one risky asset with Brownian motion dynamics and one riskless asset, it is easy to derive dynamics for the wealth of the investor and to use consumption and the share of risky asset in the portfolio as decision variables as in Merton (1971). The problem of portfolio selection with proportional costs has been a focus of recent research. Although the problem was clearly set by Magill and Constantidines as early as 1976, the mathematics of local time and reflecting diffusion processes necessary for a complete solution were relatively unknown then, precluding a solution for some years. The use of these mathematics led to several papers (Constantidines, 1986; Taksar, Klass and Assaf, 1988; a wrap up in which continuous control of consumption and singular control of transactions is allowed simultaneously in Davis and Norman, 1990); and the development of an exact solution to a particular case, by Dumas and Luciano,1991).

The case of fixed costs is fundamental in portfolio selection too. Many applications of individual portfolio selection should clearly include fixed costs of transaction. They will often combine fixed costs with the aforementioned problem of proportional costs.

There are, however, two distinct problems embedded in the fixed-costs problem: the cases of complete or partial observation. When there is full observation, the principal observes prices all along the process. In the case of stocks and many other traded assets, full observation is a good assumption. Computers allow investors today to monitor prices in real time. The mathematical problem corresponding to that situation has been solved as an "impulse control" problem by Bensoussan and Lions (1973 and 1975) and Richard (1977), without specific application to the financial problem of portfolio selection and, hence, without the specific difficulties arising there. Richard (1977) extends Bensoussan and Lions two previous papers by allowing for both fixed and proportional costs. He also allows for negative control as well as finite and infinite horizons. All cases are treated with full information.

A problem of "impulse control" arises when a Brownian motion is forced to jump back from a trigger point to a target point inside a given domain. As analyzed by Dumas (1991) , this situation must be distinguished from "optimal stopping," as in the exercise of an American option before maturity, and from "instantaneous control," where the Brownian motion is to be reflected at some trigger point. Whether in the full observation or in the partial observation case, we are dealing with an impulse-control problem here each time repeated control is considered.

Eastham and Hastings (1988) formulate the general portfolio problem (fixed and proportional costs, risky assets, Ito processes combined with jumps) in a general setting but solve it only in the specific case of no jump, one risky asset, proportional costs, and risk-neutral utility.

Grossman and Laroque (1990) approach the portfolio selection-with-fixed-costs problem when fixed costs apply to changes in consumption, instead of trading of assets. They show that consumption is not a smooth function of wealth any longer. As long as the ratio of consumption to wealth stays within a defined interval, no change in consumption is effected; but if the ratio goes out of this interval, then consumption must be adjusted to a new target level strictly inside the interval. Hence, the problem, as in the study here, is one of impulse control, not one of reflecting motion. Grossman and Laroque also show that even for small transactions costs, consumption changes can occur infrequently and hence be quite different from the continuous changes of the zero-cost case. They analyze the impact of consumption transaction costs on the demand for risky asset. All of these papers assume a full information setting.

When there is partial observation, the principal observes prices only at the time of transaction (i.e., with fixed costs for the observation itself, not only for the transaction). This situation is pertinent to the case of a guarantor monitoring the firms it guarantees only at discrete times.

Anderson and Friedman (1977) posit a quality control problem similar to ours, with costly checkups of the machine state, but with the simplified feature of having only two cost positions for the machine. Quality control is indeed isomorphic to the guarantee control problem. Suppose the cost of operating a machine depends on its performance. This condition is functionally equivalent to the cost of a guarantee depending on some assets' dynamics. Furthermore, if observing the machine's state of quality incurs initial costs even before costs of repairs, upgrades or replacements, the equivalency of this problem to the problem of guarantee control with costly audits and controls (or costly closure) becomes apparent.

Duffie and Sun (1990) consider the related problem of fixed costs incurred each time information is processed to balance a portfolio (i.e., of fixed costs with partial observation in a portfolio selection program). They show their problem to be equivalent in the case of constant relative risk aversion to a combination of a deterministic continuous-time control for consumption with a stochastic discrete time portfolio selection problem. The methodology, as well as the overall framework we use, is different. They focus on a case where, because of constant-relative risk-aversion, transaction intervals are constant. This study differs by allowing for randomly spaced partial observations.

The technology used in this study to solve this particular problem was first described by Bensoussan and Lions (1982) while solving a problem of machine quality control. In their research, Bensoussan and Lions showed both existence theorems and approximation methods for solutions. We develop the quasi-variational inequalities applied to impulse control with full and then partial observation in a similar manner to Bensoussan and Lions, but with the application to guarantees and the numerical uses of the model that will follow in that domain. This application uses more general formulations in several dimensions and provides the analysis for two cases that Bensoussan and Lions do not treat but which often occur in reality. The technology has the potential to be widely used to inform the procedures of guarantors.

Chapter 3

ELEMENTS OF OPTIMAL CONTROL

Generally speaking, the concept of control could be defined as the process of influencing the behavior of a dynamical system in order to achieve a desired goal. When the goal consists in optimizing some payoff function (or minimizing a cost function) depending on the control inputs, then we are dealing with an optimal control problem.

Optimal control problems can be found in both deterministic and stochastic settings. In the former (latter) case, the dynamics of the system to be controlled are governed by a set of ordinary (stochastic) differential equations.

1. OPTIMAL DETERMINISTIC CONTROL

In a deterministic setting such as, for example, the production planning of a factory producing n commodities, the process being controlled could be the inventory levels $y_i(s)$ at time s of any of the commodities ($i = 1, 2, ..., n$). If we further denote by $\phi_i(s)$ the production rate at time s for the i-th commodity, and by w_i the corresponding fixed demand rates (known to the planner) then the evolution of the inventory $\mathbf{y}(s) = (y_1(s), y_2(s), ..., y_n(s))'$ will be described by:

$$\frac{d}{ds}\mathbf{y}(s) = \boldsymbol{\phi}(s) - \mathbf{w}$$

with $\boldsymbol{\phi}(s) = (\ \phi_1(s),\ \phi_2(s), ...,\ \phi_n(s))'$, $\mathbf{w} = (w_1, w_2, ..., w_n)'$. If we consider a finite time interval $t \leq s \leq T$, given an initial inventory $\mathbf{y}(t) = \mathbf{x}$, the problem will be to choose the control vector representing the production rate, $\phi(s)$, in such a way as to minimize

$$\int_t^T f(\mathbf{y}(s), s)ds + \psi(\mathbf{y}(T)),$$

where T is called the *terminal time,* f is the *running* or instantaneous *cost*, and Ψ is the *terminal cost.* It is normally assumed that both f and ψ are at least continuous. However, other assumptions (such as integrability) are often considered in order to ensure that J is well-defined. Moreover, as in many other problems, the control vector $\phi(s)$ must satisfy certain constraints. In this particular case, the constraints are related to the physical capabilities of the factory and the work force and translate into upper bounds for the elements of $\phi(s)$. One such general constraint may be expressed as

$$\sum_{i=1}^{n} c_i \phi_i(s) \leq 1, \qquad \forall s.$$

and for suitable constants $c_i > 0$.

In general, under the deterministic setting, we have:

$$\frac{d}{ds}\mathbf{y}(s) = \boldsymbol{\alpha}(\mathbf{y}(s), s, \phi(s)), \qquad t \leq s \leq T,$$

with initial condition $\mathbf{x}(t) = \mathbf{x}$.

The control vector $\phi(s)$ will be a bounded, Lebesgue measurable function on $[t, T]$ with values in a closed subset $\Xi_0(t)$ of $\Re^m$, known as the *control space.* The solution of the previous differential equation is called the *state* of the system at time s, and will depend on the control vector ϕ as much as on the initial condition, even though these dependencies are not explicitly noted. For each initial data set $(t, \mathbf{x})$ it is often specified a set of admissible controls $\Xi(t, \mathbf{x}) \subset \Xi_0(t)$, and a payoff functional $J(\mathbf{x}, t; \phi)$ to be minimized, given by

$$J(\mathbf{x}, t; \phi) = \int_t^T f(\mathbf{y}(s), s, \phi(s))ds + \psi(\mathbf{y}(T)).$$

In an optimal stopping problem, where the process dynamics are monitored until exit from a closed cylindrical region $\overline{Q} = [t, T) \times \overline{O}$, that is at times $s \in (t, \tau)$, where $\tau = T \wedge t_e$ and t_e represents the exit time of $\mathbf{y}(s)$ from the closure $\overline{O}$ of an open set $O \subset \Re^n$, the payoff functional will take the form

$$J(\mathbf{x}, t; \phi) = \int_t^\tau f(\mathbf{y}(s), s, \phi(s))ds + \Upsilon(\mathbf{y}(\tau), \tau)\chi_{\tau<T} + \psi(\mathbf{y}(T))\chi_{\tau=T},$$

with χ denoting the indicator function defined as

$$\chi_{a<b} = \begin{cases} 1, & a < b \\ 0, & \text{otherwise} \end{cases}$$

The payoff function for both the fixed finite time horizon and the stopping cases can be written under the same general form:

$$J(\mathbf{x}, t; \boldsymbol{\phi}) = \int_t^{\mathcal{T}} f(\mathbf{y}(s), s, \boldsymbol{\phi}(s))ds + C(\mathbf{y}(\tau), \tau),$$

where

$$C(\mathbf{x}(\tau), \tau) = \begin{cases} \Upsilon(\mathbf{y}(\tau), \tau) & \text{if } \tau < T \\ \psi(\mathbf{y}(\tau), \tau) & \text{if } \tau = T \end{cases}$$

The control problem consists in finding the optimal control $\boldsymbol{\phi}^* \in \Xi(t, \mathbf{x})$, that is the one which, given the initial data $(t, \mathbf{x})$, yields

$$J(\mathbf{x}, t; \boldsymbol{\phi}^*) \leq J(\mathbf{x}, t; \boldsymbol{\phi}) \quad \forall \boldsymbol{\phi}(.) \in \Xi(t, \mathbf{x}).$$

The resulting minimum value of the payoff functional is called the *value function*, that is

$$u(\mathbf{x}, \mathbf{t}) = \inf_{\boldsymbol{\phi}(.) \in \Xi(t, \mathbf{x})} J(\mathbf{x}, t; \boldsymbol{\phi}) \quad \forall \ (t, \mathbf{x}) \in \overline{Q}.$$

It can be shown (see Lemma 4.2 in Fleming and Soner, 1993) that for any initial condition $(t, \mathbf{x}) \in \overline{Q}$ and for any $r \in [t, T]$ one has

$$\begin{aligned} u(\mathbf{x}, \mathbf{t}) &= \inf_{\boldsymbol{\phi}(.) \in \Xi(t, \mathbf{x})} \{ \int_t^{\tau \wedge T} f(\mathbf{y}(s), s, \boldsymbol{\phi}(s))ds + \Upsilon(\mathbf{y}(\tau), \tau)\chi_{\tau < r} \\ &\quad + u(\mathbf{y}(r), r)\chi_{\tau \geq r} \}. \end{aligned}$$

This is the central statement of the *dynamic programming principle,* that is the basis of the solution method developed by Bellman in the 50's. By assuming that the value function is continuously differentiable, it may be shown that it satisfies a nonlinear partial differential equation. Indeed, from the dynamic programming equation above, we obtain for $r = t + h$ and $h > 0$:

$$\inf_{\boldsymbol{\phi}(.) \in \Xi(t, \mathbf{x})} [\frac{1}{h} \int_t^{(t+h) \wedge T} f(\mathbf{y}(s), s, \boldsymbol{\phi}(s))ds + \frac{1}{h}\Upsilon(\mathbf{y}(\tau), \tau)\chi_{\tau < t+h} +$$

$$+ \frac{1}{h} \left(u(\mathbf{y}(t+h), t+h)\chi_{\tau \geq t+h} - u(\mathbf{x}, t) \right)] = 0.$$

It follows by letting $h \downarrow 0$:

$$-\frac{\partial}{\partial t}u(\mathbf{x},t) + \inf_{\phi(.)\in\Xi(t,\mathbf{x})}\left\{f(\mathbf{x},t,\phi(t) + \alpha(\mathbf{x},t,\phi(t))\frac{\partial}{\partial \mathbf{x}}u(\mathbf{x},t)\right\} = 0,$$

which represents a nonlinear partial differential equation of first order, often referred to as the *Dynamic Programming Equation* (*DPE*), or as the *Hamilton-Jacobi-Bellman* (*HJB*) partial differential equation. For the sake of notational simplicity, the DPE is also written as

$$-\frac{\partial}{\partial t}u(\mathbf{x},t) + H(\mathbf{x},t,D_{\mathbf{x}}u(\mathbf{x},t)) = 0,$$

where $D_{\mathbf{x}}u$ denotes the gradient of $u(.,t)$ and $H(.)$ is called the *Hamiltonian.*

The DPE is given together with appropriate boundary conditions. One such condition is for example:

$$u(\mathbf{y}(T),T) = \psi(\mathbf{y}(T)).$$

This is called a Cauchy-type condition. An additional class of conditions accompanies optimal stopping problems. Here we must also have:

$$u(\mathbf{x},t) \le \Upsilon(\mathbf{x},t)$$

The equality holds whenever $(t,\mathbf{x})$ enters what is called a *lateral boundary,* moment at which it is optimal to exit immediately from $\overline{Q}$.

2. OPTIMAL STOCHASTIC CONTROL

If in the purely deterministic case, the value function obeys to a first-order PDE, the optimal control of a piecewise deterministic process leads to a system of first-order PDE's. Now we will see that when the dynamics of the underlying process $\mathbf{y}(.)$ are stochastic, the DPE has a different form. In particular, for a *controlled diffusion* process, the DPE is a second-order parabolic PDE.

Suppose that $\mathbf{y}(s)$ satisfies a stochastic differential equation of the form

$$d\mathbf{y}(s) = \boldsymbol{\alpha}(\mathbf{y}(s),s,\boldsymbol{\phi}(s))ds + \boldsymbol{\sigma}(\mathbf{y}(s),s,\boldsymbol{\phi}(s))d\mathbf{z}(s)$$

with initial data $\mathbf{y}(t) = \mathbf{x}$ Here $\mathbf{z}$ represents a d-dimensional standard Brownian motion process defined on an interval $[t,T]$; the vector function $\boldsymbol{\alpha} = (\alpha_1, \alpha_2, ...\alpha_n)'$ is the so-called *local drift* coefficient; and $\sigma(\mathbf{y}(s), s)$ is an $n \times d$ dimensional matrix-valued function such that the *local covariance* matrix of $\mathbf{y}$ at s, $a(\mathbf{y}(s), s) = \{a_{ij}(\mathbf{y}(s), s)\}_{i,j}$, given by

$$a_{ij}(\mathbf{y}(s), s) = \sum_{l=1}^{d} \sigma_{il}(\mathbf{y}(s), s)\sigma_{jl}(\mathbf{y}(s), s)$$

are symmetric and nonnegative definite. Heuristically speaking, the dynamics of $\mathbf{y}(s)$ are now described by a dynamical system driven by an additive *white noise* with state-dependent coefficients $\sigma_{il}(\mathbf{y}(s), s)$.

The controller may be allowed to know the historic time series of $\mathbf{y}$ in order to obtain the appropriate control process $\boldsymbol{\phi}(s)$ at current time s. However, because of the Markovian nature of the problem, it should be enough to consider control processes of the form

$$\boldsymbol{\phi}(s) = \boldsymbol{\phi}(\mathbf{y}(s^-), s^-).$$

In this stochastic setting,the payoff functional to be minimized should reflect some kind of averaging. In fact, the control problem on the finite-time interval $[t, T]$ is now to minimize

$$J(\mathbf{x}, t; \boldsymbol{\phi}) = E_{\mathbf{x}} \left\{ \int_t^T f(\mathbf{y}(s), s, \boldsymbol{\phi}(s))ds + \psi(\mathbf{y}(T)) \right\}.$$

In this case, Bellman's principle of dynamic programming establishes that the value function can be expressed as the sum of the cost on $[t, t+h]$ and the minimum expected cost obtained by proceeding optimally on $(t+h, T]$ with initial data $(t+h, \mathbf{y}(t+h))$, that is:

$$u(\mathbf{x}, t) = \inf_{\boldsymbol{\phi}(.)\in\Xi(t,\mathbf{x})} E_{\mathbf{x}} \left\{ \int_t^{t+h} f(\mathbf{y}(s), s, \boldsymbol{\phi}(s))ds + u(\mathbf{y}(t+h), t+h) \right\}.$$

Define for any real-valued function φ on $[t, T] \times \Re^n$ a linear operator $\mathbf{A}$, called the *backward evolution operator*, given by:

$$\mathbf{A}\varphi(\mathbf{x}, t) = \lim_{h\downarrow 0} \left[E_{\mathbf{x}}\varphi(\mathbf{y}(t+h), t+h) - \varphi(\mathbf{x}, t)\right],$$

provided that the limit exists for each $\mathbf{x}$. It can be shown that under regularity conditions on φ, one can write (*Dynkin's formula*):

$$E_{\mathbf{x}}\varphi(\mathbf{y}(s), s) - \varphi(\mathbf{x}, t) = E_{\mathbf{x}}\left\{\int_t^s \mathbf{A}\varphi(\mathbf{y}(r), r)dr\right\}.$$

The DPE is obtained by realizing that

$$u(\mathbf{x}, t) \leq E_{\mathbf{x}}\left\{\int_t^{t+h} f(\mathbf{y}(s), s, \boldsymbol{\phi}(s))ds\right\} + E_{\mathbf{x}}\left\{u(\mathbf{y}(t+h), t+h\right\}.$$

Now, substracting $u(\mathbf{x}, t)$ from both sides, dividing by h and letting $h \downarrow 0$, one obtains:

$$\lim_{h\downarrow 0}\frac{1}{h}E_{\mathbf{x}}\left\{\int_t^{t+h} f(\mathbf{y}(s), s, \boldsymbol{\phi}(s))ds\right\} = f(\mathbf{y}(t), t, \boldsymbol{\phi}(t))$$

$$\lim_{h\downarrow 0}\frac{1}{h}\left[E_{\mathbf{x}}\left\{u(\mathbf{y}(t+h), t+h\right\} - u(\mathbf{x}, t)\right] = \mathbf{A}^{\phi}u(\mathbf{x}, t)$$

$$= \lim_{h\downarrow 0}\frac{1}{h}E_{\mathbf{x}}\left\{\int_t^{t+h} \mathbf{A}^{\phi}u(\mathbf{y}(s), s)ds\right\},$$

where $\mathbf{A}^{\phi}$ is the backward evolution operator of the controlled Markov diffusion, that is one given by:

$$\mathbf{A}^{\phi}\varphi = \frac{\partial}{\partial t}\varphi + \frac{1}{2}\sum_{i=1}^{n}\sum_{j=1}^{n} a_{ij}(\mathbf{x}, t, \boldsymbol{\phi}(\mathbf{x}))\frac{\partial^2}{\partial x_i \partial x_j}\varphi + \sum_{i=1}^{n}\alpha_i(\mathbf{x}, t, \boldsymbol{\phi}(\mathbf{x}))\frac{\partial}{\partial x_i}\varphi.$$

Therefore, for all $\boldsymbol{\phi} \in \Xi(\mathbf{x}, t)$, we have the inequality:

$$0 \leq \mathbf{A}^{\phi}u(\mathbf{x}, t) + f(\mathbf{x}, t, \boldsymbol{\phi}).$$

If $\boldsymbol{\phi}^*$ is an optimal Markov control policy then we must have:

$$u(\mathbf{x}, t) \leq E_{\mathbf{x}}\left\{\int_t^{t+h} f(\mathbf{y}^*(s), s, \boldsymbol{\phi}^*(s))ds\right\} + E_{\mathbf{x}}\left\{u(\mathbf{y}^*(t+h), t+h\right\},$$

where $\mathbf{y}^*$ represents the diffusion processobtained from $\boldsymbol{\phi}^*$. Thus

$$0 = \mathbf{A}^{\phi^*}u(\mathbf{x}, t) + f(\mathbf{x}, t, \boldsymbol{\phi}^*).$$

The DPE can be therefore expressed as

$$0 = \min_{\phi \in \Xi(t,\mathbf{x})} \left\{ \mathbf{A}^{\phi} u(\mathbf{x},t) + f(\mathbf{x},t,\phi) \right\},$$

with boundary condition

$$u(\mathbf{y}(T),T) = \psi(\mathbf{y}(T)).$$

3. STOCHASTIC IMPULSE CONTROL

In impulse control problems, the controller exerts her influence through displacements of the process done at isolated points in time. To be more precise, let (S,F,P) a probability space, where F denotes a σ-algebra of subsets on S, and P is a probability measure on F. Let $\mathbf{x}(t)$ a (generally vector-valued) random process (sequence of random variables) defined on S. Finally, let F_t represent an indexed set of sub-σ-algebras of F such that $F_t \supset F_s$ if $t > s$. A random variable θ is said to be a *stopping time* if the set $\{\theta \leq t\} \in F_t$ for each t. That is, if θ is an F_t-stopping time then whether the event $\{\theta \leq t\}$ is true could be ascertained by observing the process $\mathbf{x}(t)$ up to time t.

Now, for each $\mathbf{x}_0 \in \Re^n$ let $P_{\mathbf{x}_0}$ be the unique probability measure on (S,F) such that $\mathbf{x}(t)$ is an n-dimensional diffusion process with drift function $\boldsymbol{\alpha}$, covariance matrix $\boldsymbol{\sigma\sigma}'$, and starting value $\mathbf{x}_0$ under $P_{\mathbf{x}_0}$. Any impulse control problem can be defined in terms of a sequence of stopping times $\{\theta_0, \theta_1, ...\}$ and a sequence of vector random variables $\{\boldsymbol{\phi}_0, \boldsymbol{\phi}_1, ...\}$ such that $\boldsymbol{\phi}_i \in \Re^n$ and

$$P_{\mathbf{x}_0}(t_0 \leq \theta_0 < \theta_1 < ... \to \infty) = 1, \quad \forall \mathbf{x}_0 \in \Re^n$$
$$\boldsymbol{\phi}_n \in F_{\theta_n}, \quad \forall n = 0,1,...$$

The stopping time $.\theta_i$ should be interpreted as the i-th time at which the controller enforces a jump in the state of the process, while the elements of $\boldsymbol{\phi}_i$ represent the size of the jumps affecting the components of $\mathbf{x}(t)$ at $t = \theta_i^-$. Associated with the control policy $\{(\theta_n, \phi_n)\}$, we define the processes:

$$N(t) = \sup\{n \geq 0 : \theta_n \leq t\}, \quad t \geq 0,$$
$$\mathbf{s}(t) = \boldsymbol{\phi}_1 + ... + \boldsymbol{\phi}_{N(t)}, \quad t \geq 0,$$
$$\mathbf{y}(t) = \mathbf{x}(t) + \mathbf{s}(t), \quad t \geq 0,$$

where $\mathbf{y}(t)$ represents the controlled diffusion. Our decision variables are therefore $\{\theta_i, \phi_i\}_{i=1,N(T)}$ and $N(T)$.

As an example, consider the stochastic cash management problem in which a controller continuously monitors the content or state of a cash fund in which a certain amount of income is automatically put and out of which isolated disbursements in time are made (see Harrison, Sellke and Taylor, 1983, for details). When the content of the cash fund becomes too large (small), the controller may decide to convert some of his cash into securities (or viceversa). In such operations, she must usually pay a fixed transaction cost plus a cost proportional to the transaction size. Here, in the absence of control, the content process $y(t)$ is supposed to fluctuate as a Brownian motion with drift α and variance σ^2. The controller must at any time either increase or decrease optimally the content of the system, while keeping $y(t) \geq 0$, $\forall t$. The costs to be taken into account in this optimization problem could be modelled as: a) a fixed cost c_0 plus a proportional charge $c_1\delta$ whenever the controller effects an increase from level x to level $x + \delta$; b) a similar fixed cost c_0' plus a proportional charge $c_1'\delta$ whenever the controller decreases the content from level x to level $x - \delta$; c) inventory holding costs, incurred continuously at a rate of $\beta y(t)$, with $\beta \in (0,1)$. In this problem, we must make sure that the optimal policy is *feasible*. For an initial cash fund content of x, a policy $\{(\theta_n, \phi_n)\}$ will be feasible in an interval $[t_0, T]$ if:

$$P_x(y(t) \geq 0 \quad \forall t \in [t_0, T]) = 1, \qquad \forall x \in \Re,$$

$$E_x\left(\sum_{n=0}^{N(T)} (1 + |\phi_n|)\exp(-r\theta_n)\right) < \infty, \qquad \forall x \in \Re,$$

where r represents the cost of money, here supposed to be constant.

For such feasible policies in the fixed finite time horizon problem, the payoff function to be minimized has the form:

$$J(x, t_0, \boldsymbol{\theta}, \boldsymbol{\phi}, N(T)) = E_x\left\{\beta\int_{t_0}^{T} \exp(-rs)y(s)ds + \sum_{n=0}^{N(T)} \exp(-r\theta_n)c(\phi_n)\right\},$$

where

$$\begin{aligned} c(\phi) &= 0, \quad \text{if } \phi = 0, \\ &= c_0 + c_1\phi, \quad \text{if } \phi > 0, \\ &= c_0' + c_1'\phi, \quad \text{if } \phi < 0, \end{aligned}$$

and

$$\boldsymbol{\theta} = (\theta_0, \theta_1, ...)',$$
$$\boldsymbol{\phi} = (\phi_0, \phi_1, ...)'.$$

Chapter 4

THE MODEL

The general model we propose could be applied to a wide range of specific situations, for the control of credit risk or of any short or long position in financial guarantees. It can be used for ongoing control of guarantees (i.e., impulse control) or for optimal seizure timing (i.e., stopping time). In the real world, some situations will allow for ongoing control; others will leave only seizure as a control mean. When ongoing control is allowed, control of the guaranteed party's assets is effected by monitoring them repeatedly (either continuously if monitoring is free or at endogenously determined time intervals) and by asking for changes in asset structure regularly (i.e., asking for new collateral, equity infusion, etc.). The program will provide both the optimal times to monitor the assets and the optimal times to control them (i.e., impact them), and will also provide the amount of changes that should be requested. When only seizure is allowed for control, control of the guaranteed party's assets is effected by monitoring the assets repeatedly at endogenously determined optimal time intervals, which will undoubtedly differ from those of the previous situation, and by seizing the assets, hence freezing their value at time of seizure and realizing this value. The modified program (as will be briefly shown later) would then provide optimal timing for auditing as well as for the final time of seizure.

There are many other examples of optimal stopping/impulse control problems in the finance literature and elsewhere. The more famous applications include portfolio selection with transactions costs (Duffie and Sun,1990; Grossman and Laroque, 1990; Eastham and Hastings, 1988). Other applications include American option pricing (optimal stopping), exchange rate regulation (Flood and Garber, 1989) quality control problems (Anderson and Friedman, 1977), inventory management (Harrison,

Kreps and Taylor, 1992). This model is the first one we know of that applies the concept to credit risk problems.

1. THE UNDERLYING PROCESS BEHAVIOR

We consider one party exposed to the credit risk, also called the guarantor (short the *guarantee*) and one party that provides the credit risk, also called the guaranteed party (long the guarantee). The guarantor is thus the credit risk exposed agent while the guaranteed party is the party giving the credit risk exposure. This credit risk exposure is homogenous to the guarantor issuing a guarantee to the guaranteed party with a lifetime of T, where T might be infinite. T is the *maturity* of the bullet financial contract on which the credit risk bears. The guarantee is a bonding contract whereby the guarantor commits itself to fulfill a financial contract if the guaranteed party fails to do so. Suppose, for example, that a guarantor covers another company's liability (e.g., another company's debt, a subsidiary's debt, or a bank's deposit). The guarantor may be the government (e.g., guaranteeing deposits for a bank), a corporation (e.g., guaranteeing a subsidiary's loan), a bank (e.g., guaranteeing a letter of credit), or another financial intermediary (e.g., guaranteeing a swap contract). Each of these examples will differ in the types of assets collateralizing the guarantee and their relative amounts as well as in the maturity date of the guarantee and the seizure capacity or monitoring and control capacities of the guarantor and its costs.

Suppose formally that the guaranteed party has net assets (i.e., assets less liabilities) securing the guarantee with aggregate market value at time t given by $y(t)$. The counterparty can be one of many things: a bank, a corporation's subsidiary, a letter of credit issuer, a swap counterparty, ...; $y(t)$ is thus a stochastic process representing the value of the collateral to the guarantee (to the credit risk) at any time t. Here, the collateral can consist in a number of different assets (stocks, bonds, derivatives) or any portfolio of these.

We consider at this stage a situation in which a number of simplifying assumptions occur that make calculations easier (although many could be relaxed as discussed later):

- The *guarantor*, that is the party exposed to the credit risk, does not present credit risk itself. We thus avoid the problem of two-sided credit risk, as it appears in swaps.

- The contract considered is a bullet contract and the *credit risk*, or the *guarantee*, bears on a payment at maturity time T .

- The contract also bears on the *value of the collateral* so that the only stochastic process that affects its credit risk a this point is $y(t)$.
- We consider a situation where the value of the collateral $y(t)$ follows the following dynamics:

$$dy(t) = \alpha(y,t)dt + \sigma(y,t)dz(t)$$

where $z(t)$ is a *Wiener process.* See Merton (1971 or 1990, ch. 5) for a discussion of the use of Itô processes in the description of asset prices. As usual in finance, it makes sense to assume that the assets are of the limited-liability type so that 0 is a natural absorbing barrier. There is no mathematical necessity here in imposing 0 as an absorbing barrier.

In some situations, no specific assets are set aside as collateral for the guarantee (as in any unsecured loan). Nonetheless, there are assets affecting the value of the guarantee implicit in the contract. Otherwise there would be certainty that the guarantee will be used –as the guaranteed party has every incentive to do so– and it would not be a guarantee any more, in the sense of being an option. The guarantor bases its analysis of the guarantee on these assets' dynamics. The assets to be considered when specifying the program for a particular institution should only include those assets affecting the value of the guarantee and those that, to some degree, can be controlled. Indeed the term assets we use here may be misleading. In some cases, the value of the guarantee will not actually depend on the value of assets but rather on a different process, such as the value of a portfolio of assets and liabilities or a net worth.

It is assumed here that the process affecting the value of the guarantee is the same process as the one the guarantor can control. In some cases, no control whatsoever will be possible: Optimization is obviously not feasible then. In all other cases, even when control timing is constrained by regulation, organization problems, etc., our program will at least bring in the knowledge of what optimal control would have been. Such information may even allow the structure of the organization to change in the long term to take optimal behavior into consideration, or to gain in the awareness of the losses faced because of the lack of control. Information on optimal control, even when this optimal control cannot be implemented, could also be used in some short-fall risk compensation measures, such as hedging if the assets collateralizing the guarantee are traded. Extension of the model to situations where uncontrollable assets also affect the value of the guarantee is straightforward. Because they are uncontrollable, this kind of assets would not affect the optimal policy described later. Such assets are not considered in this analysis for the sake of the clarity of the exposition.

The model is developed with a one dimensional process. Multidimensional extensions are in principle tractable. The increase in dimensions of the system, however, soon becomes a problem for numerical solutions.

In this study, the term "asset values" refers to "market values." There are two reasons for using market values rather than book values:

(1) Refined analytical valuations of guarantees are based on market values; there is no complete theory today for determining the cost of guarantees based on book values.

(2) The dynamics of book values are more discretionary; we do not know of any process that would model them easily. As this study shows, the model must be based on an assumption of assets' dynamics.

In some cases, market values will be relatively easy to determine, notably when the assets are traded securities. In others, only book values will be available. A full audit, with the many supplementary disclosures required today and available market information otherwise should lead to reasonable estimates of market values. Each case will require separate consideration. In order to keep premiums low (a comparative advantage in the marketplace and a Pareto-superior solution in a governmental-guarantee system), optimal control is necessary to reduce all types of control errors, those apparently at the advantage of the guarantor as well as those at the advantage of the guaranteed party.

Later in the text, the use of the term "asset behavior" refers to the stochastic process that drives the dynamic path of the value of the guarantee, whether those variables are actually the prices of assets or not.

2. COST OF CREDIT RISK

We assume that the guarantor's objective is to control the evolution of its credit risk cost or the cost of the guarantee over time. We define the instantaneous evolution of the cost to the guarantor of covering the guarantee as $f(y(t), t)$. The running cost $f(y(t), t)$ depends on the level of the assets, $y(t)$. Although we do not explicit at this stage what the instantaneous cost of credit risk is, and thus derive general results, one should rely on the credit risk pricing literature as quoted above to define the objective function more precisely. We give a numerical example of analysis later relying on a simple interpretation of Merton (1974).

The expected evolution of the cost to the guarantor of the guarantee over its lifetime with no audits and no action to control the assets' behavior, and with $y(0) = x$, is therefore:

$$E_x \left\{ \int_0^T f(y(t), t) dt \right\},$$

where the expectation operator E_x carries over all possible trajectories of the process $y(t)$ starting at $y(0) = x$

The instantaneous cost of the guarantee, f, can be understood in two different perspectives. First f can be seen as a strict gross cost. In that sense, f does not include any of the revenues from the guarantee to the guarantor, such as premiums. This interpretation of f will lead to optimal behavior in those cases where the guarantor's audits and controls do not affect its revenues (i.e., the premiums requested are not impacted by its decision variables). When control affects revenues, the second interpretation of f will be needed to obtain optimal behavior. Second, treating f as a net cost (i.e., the cost of the guarantee minus the profits arising from the guarantee, the premiums) positions f as simply the negative of the profit function; minimizing the overall cost leads to the same exact program as maximizing the profit function. These two interpretations can be extended in the analysis. Nonetheless, the constraints occurring in a real-life application of the program are going to be different in the two cases.

The following analysis focuses on the second interpretation of f, that of a net cost, with the problem expressed as a cost-minimizing problem rather than a profit-maximizing one since our applications include a significant proportion of governmental applications: In that context, speaking of cost-minimization rather than profit-maximization makes more sense. For businesses, the programs are identical, i.e., the programs can just as well be described as profit-maximization programs. Because of this interpretation, f can be negative (i.e., profits positive). Because of the costs of control, however, the guarantor will not be able to drive the cost of the guarantee to -∞ (i.e., infinite profit).

3. COST OF INFORMATION

The optimal program will be different for situations in which information involves costs and when it does not. The cost of information can be divided between the cost of knowing which assets the guarantor can control and knowing how these assets behave (i.e., how much they are worth). There are some pertinent cases where there actually is almost no cost to gathering information: for example, when the collateral is well defined and consists of traded assets (e.g., a portfolio of publicly traded stocks securing a loan). The credit risk-exposed party can then obtain the market value of the assets traded at virtually no cost. This

process described by a system of stochastic differential equations is often called "stochastic control"). These techniques have been the object of numerous applications in the medical sciences, in the theory of reliability and maintenance, and in economics and finance (in problems such as determining the optimal time for buying or selling assets, the optimal lifetime consumption, investment allocation or portfolio selection, inventory problems in which a sequence of restocking decisions is made, etc.) (see Hindy, Huang and Zhu, 1997; Duffie, 1992 -chapter 8-; Bensoussan and Lions, 1982 -pp. 11-37). Here we focus on two particular forms of stochastic control: *impulse control* and *optimal stopping*, which could be seen as a particular case of the former, and viceversa (it is also possible to consider an impulse control problem as an optimal stopping problem with implicit objective or cost functions -Davis, 1993, p. 187). Impulse control techniques were first developed by Bensoussan and Lions (1973) in the context of diffusion processes, and have generated a large amount of literature as well as many applications. In the case of impulse control, dynamic programming does not lead to a single partial differential equation, but rather to a system of differential inequalities.

In this section, we briefly recall the definition of our problem and of the variables involved, we discuss two different avenues of approach to impulse control, and finally we focus on a characterization of optimality in terms of the so-called *Quasi-Variational Inequalities* (QVI's).

Chapter 5

FULL-OBSERVATION CASE

Our problem consists in deciding when to intervene on the fully-observable value of the guarantee or loan along its lifetime, T (in the partial observation case, where the guarantor can only monitor the value of the guarantee through costly audits, the problem consists also in deciding when it is best to audit). First of all, we have the value of the guarantee at time t, $y(t)$, modelled as a *diffusion* with drift and diffusion parameters, $\alpha(y(t), t)$ and $\sigma(y(t), t)$, respectively, and which depend on the underlying process. This means that when no control is exerted one has $dy(t) = \alpha(y(t), t)dt + \sigma(y(t), t)dz(t)$, where $z(t)$ denotes the standard Wiener process. When control can be exerted through the parameters, we obtain for the *controlled diffusion* the model

$$dy(t) = \alpha(y(t), e(t), t)dt + \sigma(y(t), e(t), t)dz(t)$$

where $e(t)$ is an admissible control law, that is one which is non-anticipative with respect to the process $z(t)$. Notice that unless $\sigma(y, t) \equiv 0$, realizations of the process $y(t)$ cannot force to be non-negative with this model. To make sure that $y(t)$ is always positive, we can choose a *Geometric Brownian Motion with reflection* at the barrier $y = 0$, as a model for the assets' value process. However, it is also possible to have negative values for $y(t)$ with arbitrarily small probability, by an appropriate selection of the parameters μ and σ.

In the impulse control problem, the control law will take the form of a *jump process*, which means that we can write $y(t)$ as

$$y(t) = y(0) + \int_0^t \alpha(y(s), e(s), s)ds + \int_0^t \sigma(y(s), e(s), s)dz(s) + J(t)$$

where the jump intensity at time t, $J(t)$, will depend on the value of the process just before t, $y(t^-)$. The jump process can be represented by a sequence of jump intensities , ϕ_i, and of *stopping times,* θ_i, at which the jumps affect the evolution of the process. In the optimal stopping problem, the control law will consist in seizing the assets a given instant $\tau \leq T$, which is again a *stopping time.* As a consequence, the guarantee will have an effective lifetime of $T \wedge \tau = \min(T, \tau)$.

1. THE DECISION PROCESS

The guarantor's decision criterion can be put into the form of minimizing a cost, part of which is a continuous cost at a rate $f(y(t), t)$. As the evolution of the assets' value $y(t)$ is unknown, the guarantor's criterion consists in minimizing the expected cost over the lifetime of the guarantee. If no control is exerted along this lifetime, the expected minimum cost of the credit or guarantee to the guarantor, when the initial value of the assets was $y(0) = x$,has the form

$$E_x \left(\int_0^T f(y(t), t)dt \right)$$

where the expectation operator E_x carries over all possible trajectories of the process $y(t)$ starting at $y(0) = x$, for given $\alpha(y), \sigma(y)$.

In the impulse control case, there is also a discrete cost, $c(\phi_i, \theta_i)$, which depends on the required increase in $y(t)$, that is ϕ_i, incurred at the chosen intervention or control times, θ_i. The objective function becomes therefore

$$E_x \left\{ \int_0^T f(y(t), t)dt + \sum_{i=1}^{N} c(\phi_i, \theta_i) \exp(-r\theta_i) \right\}$$

where r denotes the time value of money, or *riskless interest rate*, and N is the number of jumps required, which together with ϕ_i and θ_i, are control variables. Here, the summation term has the effect of discouraging too frequent interventions. Notice that the cost of the control at any time must be discounted back to its present value, hence the term $\exp(-r\theta_i)$.

In the optimal stopping problem, there is a single undiscounted discrete cost, $g(y(T \wedge \tau), T \wedge \tau)$, which is the cost of seizure at time τ. The cost to be minimized corresponds in this case to

$$J' = E_x \left\{ \int_0^{T \wedge t} f(y(t), t) dt + g(y(T \wedge \tau), T \wedge \tau) \exp(-r(T \wedge \tau)) \right\}$$

where the only control parameter is τ, that is, the set of admissible controls is a set of stopping times.

Given the previous variables, our impulse control problem can be stated as the minimization of the given objective function with constraints. According to *Bellman's principle of optimality*, this amounts to determining $\forall\, x, t$, the function $u(x,t)$, defined as

$$u(x,t) = \min_{\theta_i, \phi_i, N} E_x \left\{ \int_t^T f(y(s), s) ds + \sum_{i=1}^{N} c(\phi_i, \theta_i) \exp(-r(\theta_i - t)) \right\},$$

subject to

$$dy(s) = \alpha(y(s), s) ds + \sigma(y(s), s) dz(s) + \sum_{i=1}^{N} \delta(s - \theta_i) \phi_i,$$

with $y(t) = x$, and where $\delta(s)$ is the delta of Dirac at time s.We must bear in mind that all control variables, $\{\theta_i, \phi_i, N\}_i$, depend on T.

Similarly, our optimal stopping problem amounts at determining:

$$\min_{\tau} E_x \left\{ \int_t^{T \wedge \tau} f(y(t), t) dt + g(y(T \wedge \tau), T \wedge \tau) \exp(-r(T \wedge \tau - t)) \right\},$$

subject to

$$dy(s) = \alpha(y(s), s) ds + \sigma(y(s), s) dz(s) + d\eta(s),$$

where $\eta(s)$ is the jump process associated with the particular stopping problem.

In both problems, we decide to exert some 'discrete control' over the continuous process $y(t)$ as soon as it hits a target set. This is accomplished by introducing one or more jumps at some optimally-determined stopping times.

2. "SINGLE JUMP" OPERATOR APPROACH

The *single-jump operator approach* seems to be the most direct method, but it usually requires a substantial amount of computation.

Let $u(x) = u(x,0)$, that is, the minimum cost of the guarantee or credit to the guarantor at time 0, knowing that optimal control will be enforced later on and that $y(0) = x$. In the more general case of an impulse control problem, this method, proposed in Davis (1993), consists in the iteration of an operator, L, as follows

$$Lu(x) = \inf_{t<T} E_x \left\{ \int_0^t f(y(s),s)ds + \exp(-rt)\mathbf{M}u(y) \right\}$$

where $y(t)$ is modelled as a *piecewise-deterministic Markov process* (essentially, a process which is deterministic in-between stochastic jumps, where the jump mechanism is determined by two functions, namely, the jump rate and a transition probability measure), x is the initial state of the process and $\mathbf{M}$ is a nonlinear operator having the form

$$\mathbf{M}u(y) = \inf_z c(z-y) + u(z)$$

Under certain regularity conditions (Davis, 1993), L gives rise to a monotonically decreasing sequence $u_n(x)$, given by $u_{n+1}(x) = Lu_n(x)$, and whose generic term represents the cost of the guarantee to the guarantor when at most n interventions are allowed. This sequence admits an infimum limit value, u_∞, the fixed-point operator of L, whose existence and unicity is shown in Davis (1993). This limit value is called the *value function,* and represents the minimum cost of the guarantee to the guarantor obtained by requiring increases in the assets' value at a discrete set of optimally chosen stopping times. These intervention times cannot be determined along with the different terms of the guarantee cost sequence, v_n, but need be obtained as those instants at which the present value of the guarantee equals the value function. In the optimal stopping problem, the procedure is similar. Once the value function is determined via iteration of a monotone nonlinear operator, the optimal stopping time is obtained as the instant at which the cost function equals the value function (see Davis, 1993, for details).

3. QVI APPROACH

Let $u(x,t)$ denote the minimum expected cost to the guarantor at time t, knowing that optimal control will be enforced after t,and that $y(t) = x$. The *Quasi-Variational Inequalities* (QVI) characterization can be derived as follows. At any time t, the guarantor can either effect a control on the assets' dynamics or let the process evolve. A control will only be effected at t, if the cost of the control plus the future minimum cost of the future controlled guarantee, is exactly the minimum cost of the guarantee or credit at t. Formally, if

$$u(y(t), t) = \mathbf{M}u(y(t), t)$$

where $\mathbf{M}$ is a monotone nonlinear operator defined as

$$\mathbf{M}u(y(t), t) = \min_{\phi} \left\{ c(\phi, t) + u(y(t^{-}) + \phi, t) \right\} \text{ (impulse control)}$$
$$\mathbf{M}u(y(t), t) = g(y(t), t) \text{ (optimal stopping)}$$

If on the contrary, one has $u(y(t), t) < \mathbf{M}u(y(t), t)$. the guarantor would better not effect a control at t, and the process $y(t)$ should be left to evolve freely for at least a short time interval. By letting τ denote the amplitude of this interval. The following equation then holds

$$u(y(t), t) = E_x \left\{ \int_t^{t+\tau} f(y(s), s)ds + u(y(t+\tau), t+\tau) \exp(-r\tau) \right\}.$$

A Taylor series expansion is then possible. If, for instance, we expand $u(y(t+\tau), t+\tau)$ around t, we obtain

$$u(y(t+\tau), t+\tau) = u(y, t) + \tau u_t + \tau u_y \alpha(y(t), t) + 1/2 \tau u_{yy} \sigma^2(y(t), t) + o(\tau)$$

where u_x and u_{xx} denote, respectively, the first and second partial derivative of $u(y(t+\tau), t+\tau)$ with respect to x, evaluated at $\tau = 0$. Now, by the Mean Value Theorem, we have that

$$\int_t^{t+\tau} f(y(s), s)ds = f(y(h), h)\tau$$

for some $h \in [t, t+\tau]$. It follows that

$$u(y(t), t) \leq \min E_x \left\{ f(y(h), h)\tau + u(y(t+\tau), t+\tau) \exp(-r\tau) \right\}.$$

That is,

$$u(y(t),t) \leq f(y(h),h)\tau + (1-r\tau)u(y(t),t) + \tau u_t + \tau u_y \alpha(y(t),t) + \\ + 1/2\tau u_{yy}\sigma^2(y(t),t) + o(\tau)$$

Rearranging terms and taking limits as τ tends towards 0, we obtain our first quasi-variational inequality:

$$ru(y(t),t) - u_t - u_y\alpha(y(t),t) - 1/2u_{yy}\sigma^2(y(t),t) - f(y(t),t) \leq 0$$

Or more briefly

$$\mathbf{A}u - f \leq 0$$

where $\mathbf{A}$ is a second-order differential operator having the form

$$Au(y,t) \equiv -\partial u(y,t)/\partial t - \alpha(y,t)\partial u(y,t)/\partial y - 1/2\sigma^2(y,t)\partial^2 u(y,t)/\partial y^2 + r.$$

On the other hand, from the previous discussion, we must have

$$u - \mathbf{M}u \leq 0,$$

which is our second quasi-variational inequality. The system formed by these two inequalities can be solved by imposing sufficient additional conditions. These are the following. First, since the controller has to either apply or not a control, the strict-sense inequalities cannot hold simultaneously, that is

$$(\mathbf{A}u - f)(u - \mathbf{M}u) = 0.$$

Finally we obtain our system of QVI's, which is merely one of stochastic partial differential equations (SPDE's) having the form

$$\begin{aligned} \mathbf{A}u - f &\leqslant 0 \\ u - \mathbf{M}u &\leqslant 0 \\ (\mathbf{A}u - f)(u - \mathbf{M}u) &= 0 \\ u(y,T) = 0, \ \forall y &\in S. \end{aligned}$$

Thus an alternative solution approach begins by characterizing the value function of the impulse control or optimal stopping problem as the unique solution of a system of QVI's. This approach was suggested by Bensoussan and Lions (Bensoussan and Lions, 1982) in a theoretical context. One purpose of this paper is to present an attempt to find directly the numerical solution of the system of QVI's into which the problem of controlling collateral credit risk has been previously formulated. Recall that $u(x,t)$ is the function representing the minimum expected cost to the guarantor of the guarantee or credit at time t assuming that optimal control is enforced after this time instant, and that $y(t) = x$. This function could be defined recursively as :

$$u(x,t) = \begin{cases} E_x \left\{ \int_t^{t+\tau} f(y(s),s)ds + u(y(t+\tau),t+\tau)\exp(-r\tau) \right\}, \\ \qquad \text{if no control in } (t,t+\tau) \\ \\ \min_\phi \left\{ c(\phi,t) + u(y(t^-)+\phi,t) \right\}, \\ \qquad \text{if control is needed at } t. \end{cases}$$

The problem is to find the minimum cost of the credit to the guarantor, $u(x,t)$, for any value $x \in S$ of the process at time t, as well as the controls $\{\phi_i, \theta_i\}_i$ achieving that value. This can be solved in two steps. First of all, we obtain the function $u(x,t)$ $\forall t, \forall x \in S$,and then we determine the *continuation set,* $\mathcal{C}_{ic}$, defined as

$$\mathcal{C}_{ic} = \left\{ (y,t); u(y,t) < \min_{\phi \neq 0} \left\{ c(\phi,t) + u(y(t^-)+\phi,t) \right\} \right\}$$

Secondly, once $\mathcal{C}_{ic}$ has been obtained, the guarantor can, at any time, ascertain whether $y(t)$, the current value of the assets is such that $(y(t),t) \in \mathcal{C}_{ic}$. If this is the case, the guarantor let the process evolve naturally (without control). Otherwise, she requests an amount of collateral equal to $\arg\min_\phi \left\{ c(\phi,t) + u(y(t^-)+\phi,t) \right\}$. By proceeding in this way, we can obtain the set of controls needed. In the optimal stopping case, the continuation set has the form

$$\mathcal{C}_{os} = \left\{ (y,t); u(y,t) < g(y,t) \right\}.$$

In this case, the controller or guarantor seizes the assets affecting the credit or guarantee as soon as $(y(t),t)$ is in $\mathcal{C}_{os}$.

It can be shown that the solution to this set of inequalities is the unique fixed-point of a concave and monotonically increasing operator, **T**, defined as follows

$$
\begin{aligned}
z &\leq E_x \left\{ \int_0^t \exp(-\alpha s)\Phi(s) f ds + \exp(-\alpha t)\Phi(t) z \right\} \\
z &\leq \mathbf{M}v \\
\mathbf{T}v &= z
\end{aligned}
$$

Φ denoting a semigroup operator, that is an operator with the following properties:

$$
\begin{aligned}
\Phi(0) &= \mathbf{I} \\
\Phi(t+s) &= \Phi(t)\Phi(s) \\
\Phi(t)z &= z(y(t), t) \\
\Phi(s)f &= E_x \left(f(y(s), s) \right),
\end{aligned}
$$

with **I** denoting the identity operator, and with initial condition $y(0) = x$.

In other words, we must find the unique fixed-point of the operator mapping v to z, which, as we said, corresponds to the solution u of

$$
\begin{aligned}
\mathbf{A}u - f &\leq 0 \\
u - \mathbf{M}u &\leq 0 \\
(\mathbf{A}u - f)(u - \mathbf{M}u) &= 0
\end{aligned}
$$

with boundary condition $u(y, T) = 0$, $\forall y \in S$.

Chapter 6

PARTIAL OBSERVATION CASE

In many situations, the guarantor or creditor will need to invest some amount of money or management time in order to obtain the information on the current position of the guaranteed party's assets. This situation will change the program in two different dimensions: (1) the dynamics of the controlled assets' value, $y(t)$, and (2) the costs faced by the guarantor. Because of these auditing costs, the guarantor will be led to observe the assets' value process, $y(t)$, at only some optimally selected time instants. Clearly, the programs derived under full observation can no longer be used here. In the following, we approach the optimal impulse control problem in a partial observation context by analogy with the full observation programs. For this, we assume that the guaranteed party has a passive behavior, thus avoiding the need to consider the game-theoretical framework in which to find the general solution to this problem. Also, we will only discuss the solution to the impulse control problem, since for the optimal stopping problem the solution can be deducted from the former, by replacing the control costs with the stopping cost.

1. THE DECISION PROCESS

Ideally, the guarantor or creditor would like to have information on the assets' value all the time, but because of the auditing costs, the optimal decision at t may be only one of these:

1. Let the process $y(t)$ evolve according to its uncontrolled dynamics.

2. Make a costly audit with cost $c_0(t)$, without imposing new collateral infusion.

3. Impose a control with no auditing, that brings the level of the assets to a new starting value of $Y(t)$ at time t, at a cost of $c_1(Y(t), t)$.

According with the previous courses of action, the guarantor's goal will be to minimize the objective function given by

$$J = E_x\{ \int_0^T f(y(t),t)dt + \sum_{i=1}^{N} \left[\chi_{\gamma_i=1}\, c_0(\theta_i) \exp(-r\theta_i) + \chi_{\gamma_i=0}\, c_1(Y_i,\theta_i) \exp(-r\theta_i) \right] \}$$

with respect to the variables $\{N, \theta_i, \gamma_i, Y_i\}$, where θ_i are the *intervention times* (i.e. times at which the process is either audited or impulse-controlled), and γ_i is a control variable, taking the values 1 or 0, according to

$$\gamma_i = \begin{cases} 1, & \text{if an audit is made at } \theta_i \\ 0, & \text{if a control is imposed at } \theta_i \end{cases}$$

The dynamics of the controlled process can be written as

$$dy(t) = \alpha(y(t),t)dt + \sigma(y(t),t)dz(t) + \sum_{i=1}^{N}(1-\gamma_i)\ \delta(t-\theta_i)\left[Y_i - y(\theta_i^-)\right],$$

that is, the process evolves naturally except at times in which an impulse is applied ($\gamma_i = 0$). In this case, the impulse size will be $Y_i - y(\theta_i^-)$, where $y(\theta_i^-)$ denotes the uncontrolled level of the assets just before $t = \theta_i$, which may not be known to the guarantor.

In the optimal stopping problem, we want to minimize with respect to the stopping time variable τ, the functional

$$J' = E_x\{ \int_0^{T\wedge\tau} f(y(t),t)dt + \sum_{\substack{i \\ \theta_i < T\wedge\tau}} c_0(\theta_i) \exp(-r\theta_i) + g(y(T\wedge\tau), T\wedge\tau) \exp(-r(T\wedge\tau))\}$$

where the stopping cost $g(y(T\wedge\tau), T\wedge\tau)$ replaces the control costs, and where $\{\theta_i\}_i$ are just the auditing times.

According to Bellman's principle of optimality, the minimization of J above amounts at determining $\forall\ x, t$ the function

$$u(x,t) = \min_{N,\theta_i,\gamma_i,Y_i} E_x\{\int_t^T f(y(s),s)ds +$$
$$+\sum_{i=1}^{N}\left[\chi_{\gamma_i=1}\, c_0(\theta_i) + \chi_{\gamma_i=0}\, c_1(Y_i,\theta_i)\right]\exp(-r(\theta_i - t))\}.$$

However, the process $y(t)$ is not observable except at auditing times. Thus to apply the dynamic programming reasoning, we must replace this process by one or more that can be observed at any time.

Following Bensoussan and Lions (1982), now we show that, mediating a change of variables, the partial observation problem can be framed as a full-observation one but with different observable processes. Indeed, all the information accessible to the guarantor at any time instant can be summarized in:

a) The time elapsed since the last intervention time, given by

$$\xi(s) = \begin{cases} s-t, & t \le s < \theta_1 \\ s-\theta_i, & \theta_i \le s < \theta_{i+1} \end{cases}$$

b) The value of the assets at the last observation time, given by

$$\lambda(s) = \begin{cases} y(t) = x, \; t \le s < \theta_1 \\ y(\theta_i), \; \theta_i \le s < \theta_{i+1}. \end{cases}$$

2. QVI APPROACH

Our problem consists in finding the set of values for $\{N, \theta_i, \gamma_i, Y_i\}$ that minimize J subject to the dynamics of the controlled process, $y(t)$, as written before. First notice that we can split

$$E_x\left\{\int_t^T f(y(s),s)ds\right\}$$

into

$$E_x\left\{\int_t^{\theta_1} f(y(s),s)ds + \sum_{i=1}^{n}\int_{\theta_i}^{\theta_{i+1}} f(y(s),s)ds\right\}$$

where we define $n = \sup(i)$ and $T = \theta_{n+1}$. Let us now define the operator

$$\Omega(s, s-t)a(x) = E_x\left(a(y\left[x;t\right](s))\right)$$

with $y\left[x;t\right](s)$ denoting the solution at s of the dynamics of y, with initial condition $y(t) = x$. Accordingly, $\Omega(s, s-t)$ is the operator that yields the expectation of the argument function's value at time s, when the variable in the function, y, evolves from a starting value of x at t, during a period of length $s - t$. Therefore we can write

$$E_x\left\{\int_t^{\theta_1} f(y(s), s)ds\right\} = \int_t^{\theta_1} \Omega(s, s)f(x, s)ds$$

$$E_x\left\{\int_{\theta_i}^{\theta_{1+1}} f(y(s), s)ds\right\} = \int_{\theta_i}^{\theta_{1+1}} \Omega(s, s-\theta_i)f(y(\theta_i), s)ds$$

and

$$E_x\left\{\int_t^{T} f(y(s), s)ds\right\} = \int_t^{T} \Omega(s, \xi(s))f(\lambda(s), s)ds$$

Finally, we have

$$\begin{aligned} J &= \int_t^T \Omega(s, \xi(s))f(\lambda(s), s)ds + \\ &\quad + E_x\left\{\sum_{i=1}^{N}\left[\chi_{\gamma_i=1}\; c_0(\theta_i) + \chi_{\gamma_i=0}\; c_1(Y_i, \theta_i)\right] \exp(-r(\theta_i - t))\right\} \end{aligned}$$

Now recall that in the full-observation case, the minimum cost of the guarantee at time t, when optimal control is assumed afterwards, is given by

$$u(x,t) = \min_{\theta_i \phi_i, N} E_x\left\{\int_t^T f(y(s), s)ds + \sum_{i=1}^{N} c(\phi_i, \theta_i)\exp(-r(\theta_i - t))\right\}.$$

This was obtained as the solution of a system of QVI's. In the partial-observation case, this minimum expected cost has the form

$$u(\lambda(t), \xi(t), t) = \min_{\theta_i \phi_i, N, \zeta_i, \gamma_i} \left\{ \int_t^T \Omega(s, \xi(s)) f(\lambda(s), s) ds + \right\}$$
$$+ E_x \left(\sum_{i=1}^{N} \left[\chi_{\gamma_i = 1} \; c_0(\theta_i) + \chi_{\gamma_i = 0} \; c_1(Y_i, \theta_i) \right] \exp(-r(\theta_i - t)) \right)$$

which can be shown to be the solution of the system of QVI's:

$$\mathbf{A}^{\dagger} u - F \leq 0$$
$$u - \mathbf{M}^{\dagger} u \leq 0$$

with conditions

$$(\mathbf{A}^{\dagger} u - F)(u - \mathbf{M}^{\dagger} u) = 0$$
$$u(\lambda, \xi, T) = 0, \; \forall \; (\lambda, \xi) \in \Re^{+} \times \Re^{+}$$

and with

$$F(\lambda(t), \xi(t), t) = \Omega(t, \xi(t)) f(\lambda(t), t)$$
$$\mathbf{A}^{\dagger} u(\lambda(t), \xi(t), t) = r - \partial u(\lambda(t), \xi(t), t)/\partial t - \partial u(\lambda(t), \xi(t), t)/\partial \xi(t)$$
$$\mathbf{M}^{\dagger} u(\lambda, \xi, t) = \min \left\{ \begin{array}{c} c_0(t) + \Omega(t, \xi(t^-)) \; u(\lambda(t^-), 0, t) \\ \\ \inf_Y \left[c_1(Y, t) \; + u(Y, 0, t) \right] \end{array} \right\}.$$

To show this, we first notice that the expected cost of making an audit at time t is

$$c_0(t) + \Omega(t, \xi(t^-)) \; u(\lambda(t^-), 0, t).$$

To decide whether to audit or not, the guarantor compares the minimum expected cost today to this cost. She may also conclude that the situation can be expected to be sufficiently bad as to impose immediate control without prior auditing. If the assets are then forced to a new level, $Y(t)$, the resulting minimum cost would be

$$c_1(Y(t), t) \; + u(Y(t), 0, t)$$

Of course, the guarantor chooses the level $Y(t)$ so as to minimize this cost, that is,

$$Y(t) = \inf_{Y} \left[c_1(Y,t) \ + u(Y,0,t) \right].$$

Therefore, one must have

$$u(\lambda(t),\xi(t),t) \leq \min \left\{ \begin{array}{c} c_0(t) + \Omega(t,\xi(t^-)) \ u(\lambda(t^-),0,t) \\ \\ \inf_Y \left[c_1(Y,t) \ + u(Y,0,t) \right] \end{array} \right\}$$
$$= \mathbf{M}^\dagger \ u(\lambda(t),\xi(t),t).$$

If exercising a control or making an audit at t was not optimal then

$$\begin{aligned} u(\lambda(t),\xi(t),t) \quad &= \quad \int_t^{t+\tau} \Omega(s,\xi(s)) f(\lambda(s),s) \\ &+ u(\lambda(t+\tau),\xi(t+\tau),t+\tau) \exp(-r\tau). \end{aligned}$$

Now letting

$$F(\lambda(s),\xi(s),s) = \Omega(s,\xi(s)) f(\lambda(s),s),$$

and noting that

$$\begin{aligned} \lambda(t+\tau) &= \lambda(t) \\ \xi(t+\tau) &= \xi(t) + \tau, \end{aligned}$$

a Taylor series expansion of $u(\lambda(t+\tau),\xi(t+\tau),t+\tau)$ around t to the first order, obtains

$$u(\lambda(t+\tau),\xi(t+\tau),t+\tau) = u(\lambda(t),\xi(t),t) + \tau u_t(\lambda(t),\xi(t),t) +$$

$$+\tau u_\xi(\lambda(t),\xi(t),t) + o(\tau).$$

On the other hand, by the Mean Value Theorem, we have

$$\int_t^{t+\tau} F(\lambda(s),\xi(s),s) ds = F(\lambda(k),\xi(k),k)$$

for some $k \in (t, t+\tau)$. By substituting into the expression of $u(\lambda(t), \xi(t), t)$ given in terms of $u(\lambda(t+\tau), \xi(t+\tau), t+\tau)$, we get the first QVI:

$$ru(\lambda(t), \xi(t), t) - u_t(\lambda(t), \xi(t), t) - u_\xi(\lambda(t), \xi(t), t) \leq F(\lambda(t), \xi(t), t)$$

or more compactly,

$$\mathbf{A}^\dagger u(\lambda(t), \xi(t), t) - F(\lambda(t), \xi(t), t) \leq 0.$$

Once $u(\lambda(t), \xi(t), t)$ has been determined by solving the system of QVI's, we define the continuation set as

$$\mathcal{C}_{ic}^\dagger = \left\{ (\lambda, \xi, t) : u(\lambda, \xi, t) < \mathbf{M}^\dagger u(\lambda, \xi, t) \right\}.$$

Suppose $u(\lambda(\theta_1^-), \xi(\theta_1^-), \theta_1^-) = \mathbf{M}^\dagger u(\lambda(\theta_1^-), \xi(\theta_1^-), \theta_1^-)$. Then if

$$\mathbf{M}^\dagger u(\lambda(\theta_1^-), \xi(\theta_1^-), \theta_1^-) = c_0(\theta_1) + \Omega(\theta_1, \xi(\theta_1^-))\; u(\lambda(\theta_1^-), 0, \theta_1)$$

we must make an audit and observe $y(\theta_1)$,value on which future decisions will be based. Otherwise, we should require additional collateral infusion without prior auditing. Now, at $t = \theta_1$, we set $\lambda(\theta_1) = y(\theta_1), \xi(\theta_1) = 0$. The guarantor then lets the process evolve by itself until it hits the boundary

$$u(\lambda(\theta_2^-), \xi(\theta_2^-), \theta_2^-) = \mathbf{M}^\dagger u(\lambda(\theta_2^-), \xi(\theta_2^-), \theta_2^-)$$

at $t = \theta_2$. The procedure continues in this way until $t = T$. Notice that at any time t, the only information needed is $\lambda(t^-)$ and $\xi(t^-)$,which are readily available.

Chapter 7

NUMERICAL APPROACHES

In the application of optimal control techniques to management problem, numerical methods are often used to implement the solution due to the intrinsic complexity of the problems, which prevents from obtaining qualitative information on the system's behavior. In spite of the large amount of research done in computational methods for optimal control, the part concerning management problems is rather small.

Essentially, most numerical analysis techniques in optimal control fall in one of the following categories:

a) Numerical methods based on dynamic programming.

b) Indirect methods.

c) Direct methods.

d) Simulation.

In a), a dynamic programming algorithm solves the optimal control problem, where no boundary conditions are to be satisfied by the process. In b), the value function is not sought directly, but rather the parameters are varied until all constraints are satisfied. Contrary to b), in c) the value function is searched while satisfying the constraints of the problem. Finally, in d) the problem is "run" rather than solved.

In this section, we describe the numerical approach which we followed to solve the system of QVI's in the full observation case. For that we need an appropriate finite-difference approximation scheme. All numerical schemes first proceed by discretizing both the state-space, S,and the time interval $[0, T]$. Our continuous-time process for the assets, $y(t)$, must then be approximated by a sequence of *Markov chains.* In this way, the original optimization problem is approximated by a sequence of discrete parameter optimization problems. The DPE of the discrete-control problem can be obtained naturally as a finite-difference approximation to the continuous-time DPE. For example, consider a *time step* $\Delta t > 0$ and a *spatial step* $\Delta y > 0$. The approximating controlled discrete-time Markov chain will have as state space the one-dimensional lattice:

$$S_{x,\Delta y} = \{y = x + j\Delta y, \quad j = ... - 2, -1, 0, 1, 2, ...\}.$$

The approximating Markov chain sequence must be such that, as the stepsizes converge to zero, the mean and mean-square change per step converges to the local drift and the local variance, respectively of the continuous-time process.

If we further require that the discretized state-space $S_{x,\Delta y}$ (where Δy stands for the state-space stepsize) be finite-dimensional then we may need to consider a reflecting boundary for the finite state Markov chain, since otherwise the latter could leave $S_{x,\Delta y}$. This reflecting boundary would project the approximating chain back into $S_{x,\Delta y}$ as soon as its sample paths leave it. Moreover, this must be done in a way that is consistent with the dynamics of the process. Obviously, considering a reflecting boundary in our problem translates into imposing an additional boundary condition, which we obviate by restricting our analysis to time intervals where the chain remains within $S_{x,\Delta y}$.

From here, we must compute the sort of average effected by the expectation operator in the payoff functional. This can be done in different ways. One way is to specify the dynamics of the controlled Markov chain by giving the one-step transition probabilities to nearest-neighbor states, $\{p(y + \Delta y, \phi), p(y - \Delta y, \phi), p(y, \phi)\}$, as done for instance in Fleming and Soner (1992), and in Hyndy, Huang and Zhu (1997).

The DPE for the value function of the discretized problem, $u_{\Delta y}(x, t)$, can be shown to be (see Fleming and Soner, 1992, p. 368):

$$u_{\Delta y}(x, t) = \min_{\phi}\{p(y + \Delta y, \phi)u_{\Delta y}(x + \Delta y, t + \Delta t) +$$

$$+p(y - \Delta y, \phi)u_{\Delta y}(y - \Delta y, t + \Delta t) + p(y, \phi)u_{\Delta y}(y, t + \Delta t) + \Delta y f(x, t, \phi)\},$$

which can be solved by either implicit or explicit finite-difference schemes.

Another way of getting rid of the expectation operator E_x in the payoff functional is by simulation, as it is done in the "single-jump operator approach", briefly described in a previous section.

Finally, as shown in another section, our choice was to exploit the particular moment properties of Brownian motions, and replace the averaging operation by an expression involving temporal and spatial derivatives of the value function. In our impulse control problem, the unconstrained discretized problem becomes that of minimizing the following payoff::

$$
\begin{aligned}
J_{\Delta y,\Delta t}(x, t_k, \boldsymbol{\phi}, \boldsymbol{\theta}, N(t_M)) &= E_x\{\Delta y \sum_{i=t_k}^{t_M} \exp(-ri) f(y_i, i) \\
&\quad + \sum_{n=0}^{N(t_M)} \exp(-r\theta_i) c(\phi_i)\},
\end{aligned}
$$

subject to the dynamics of a controlled diffusion, and where $t_k - t_{k-1} = \Delta t$, $t_M = T$.

The presence of nonlinear constraints or *obstacles* causes additional difficulties in finding optimal solutions. Numerical methods dealing with obstacles can be classified as *feasible*, if the entire sequence of points generated in the iterative process satisfies these constraints (that is, they are *admissible*), and not feasible, if there are intermediate points in the iteration that are not admissible. Basically, we can talk of two different techniques. On the one hand, *projection methods*, where algorithms for the unconstrained case are adapted by considering projection operators that force the points to belong to the admissible region (that is, to satisfy the set of constraints). However, when the constraints are very nonlinear, the projection of a point could be a difficult task. On the other hand, *penalty* or *barrier methods*, where a sequence of unconstrained problems is solved, the solution of which converges towards the solution of the constrained problem. To do this, constraints are usually forgotten by adding to the payoff functional a penalty term that is zero when the point satisfies the constraints and very large otherwise. This creates a barrier preventing the point from leaving the set of constraints. The difficulty of this technique lies in finding an admissible starting point.

In our particular impulse control problem, a scheme falling within the latter class of methods is the so-called *regularization* or *penalization approach* (Bensoussan and Lions, 1973). Under this approach, given a positive real number ξ, we must look for the solution to the equation

$$\mathbf{A}u_\xi + \frac{1}{\xi}(u_\xi - \mathbf{M}u_\xi)^+ = f$$

with boundary condition $u(y, T) = 0$, $\forall y \in S_{x,\Delta y}$, and where $(a)^+ = \max(a, 0)$.

The penalized numerical scheme based on iterating the solution to

$$\mathbf{A}u_\xi^{(k)} + \frac{1}{\xi}(u_\xi^{(k)} - \mathbf{M}u_\xi^{(k-1)})^+ = f$$

on k, leads to an approximation from above to the expected minimum cost solution, u, that tends towards the latter as $\xi \to 0$ (see Theorem 3, Bensoussan and Lions, 1973). The convergence of a decreasing iterative numerical scheme in this approach to find the optimal solution is established in Glowinski *et al.* (lemma 1.2, page 313).

Here we used a projection method to find out the solution to our constrained optimization problem. Our technique splits into two stages:

- (a) The resolution of a variational equality by solving backwards the stochastic partial differential equation

$$\mathbf{A}(t_j)u^{(k)}(y_i,t_j) + \frac{1}{\xi_k}\left(u^{(k)}(y_i,t_j) - \mathbf{M}u^{(k-1)}(y_i,t_j)\right)^+ - f(y_i,t_j) = 0,$$

  for a sequence of positive real numbers ξ_k such that $\lim_{k\to\infty}\xi_k = 0$, and for each j on a grid $0 \le i \le n-1, 1 \le j \le m$, where $y_i < y_{i+l}$, for $l > 0$, and where $t_1 = 0$ and $t_m = T$. This amounts to solving backwards the system of linear equations

$$\overline{\mathbf{A}}^{(k)}(t_j)U_j^{(k)} = D_j^{(k)},$$

  for each j on the grid $0 \le i \le n-1, : 0 < j \le m-1$, and where $\overline{\mathbf{A}}$ is the regularized second-order differential operator of the parabolic PDE

$$\overline{\mathbf{A}}_j = \begin{pmatrix} b_{0,j} & c_{0,j} & 0 & 0 & \cdots & 0 \\ a_{1,j} & b_{1,j} & c_{1,j} & 0 & \cdots & 0 \\ \vdots & \vdots & \vdots & \vdots & \cdots & \vdots \\ 0 & \cdots & 0 & a_{n-2,j} & b_{n-2,j} & c_{n-2,j} \\ 0 & \cdots & 0 & 0 & b_{n-1,j} & c_{n-1,j} \end{pmatrix}$$

  and, following the *Crank-Nicholson* discretization scheme

$$
\begin{aligned}
U_j^{(k)} &= \left(u^{(k)}(y_0, t_j), \cdots, u^{(k)}(y_{n-1}, t_j)\right)', \\
D_j^{(k)} &= \left(d^{(k)}(y_0, t_j), \cdots, d^{(k)}(y_{n-1}, t_j)\right)', \\
d^{(k)}(y_0, t_j) &= b_{1,j}^{(k)} u^{(k)}(y_0, t_j) + c_{1,j} u^{(k)}(y_1, t_j), \\
d^{(k)}(y_{n-1}, t_j) &= a_{n-1,j} u^{(k)}(y_{n-2}, t_j) + b_{n-1,j}^{(k)} u^{(k)}(y_{n-1}, t_j), \\
d^{(k)}(y_i, t_j) &= -a_{i,j} u^{(k)}(y_{i-1}, t_{j+1}) + \beta_{i,j} u^{(k)}(y_i, t_{j+1}) && (7.1) \\
&\quad - c_{i,j} u^{(k)}(y_{i+1}, t_{j+1}) + f(y_i, t_j) && (7.2) \\
&\quad + \frac{1}{\xi_k} \mathbf{1}_{\mathbf{M}u^{(k)}(y_i,t_j) < u^{(k)}(y_i,t_j)}, && (7.3) \\
a_{i,j} &= -\alpha(y_i, t_j)/4\Delta y + \sigma^2(y_i, t_j)/4(\Delta y)^2, \\
b_{i,j}^{(k)} &= -1/\Delta t - \sigma^2(y_i, t_j)/2\Delta y^2 \\
&\quad + \frac{1}{\xi_k} \mathbf{1}_{\mathbf{M}u^{(k)}(y_i,t_j) < u^{(k)}(y_i,t_j)}, && (7.4) \\
c_{i,j} &= \alpha(y_i, t_j)/4\Delta y + \sigma^2(y_i, t_j)/4\Delta y^2 \\
\beta_{i,j} &= -1/\Delta t + \sigma^2(y_i, t_j)/2\Delta y^2
\end{aligned}
$$

and $\Delta y = y_i - y_{i-1}$, $\Delta t = t_j - t_{j-1}$, and with:

$$
\mathbf{1}_{\mathbf{M}u^{(k)}(y_i,t_j) < u^{(k)}(y_i,t_j)} = \begin{cases} 1, & \text{if } \mathbf{M}u^{(k)}(y_i, t_j) < u^{(k)}(y_i, t_j) \\ & 0, \text{ otherwise.} \end{cases}
$$

- (b) The projection of the solution on a constraint set defined by the obstacle. Recall that the obstacle has, in our case, the form

$$\mathbf{M}u(y(t), t) = \inf_\phi [c(\phi) + u(y(t^-) + \phi, t)].$$

The projection is done by selecting at each instant j the solution

$$u_{i,j}^* = \min\{u_{i,j}, \mathbf{M}u_{i,j}\}$$

where $u_{i,j} = u(y_i, t_j)$ is the solution obtained in (a), and $\mathbf{M}u_{i,j} = \inf_\phi [c(\phi) + u(y_i + \phi, t_j)]$.

The method requires a few iterations. At each iteration the two operations (a) and (b) are carried out for each j. The number of iterations required for convergence is usually small (between 5 and 10 iterations were enough for convergence in the experiments). Some convergence results for this iterative method can be found in Glowinski *et al.* (1981), pages 181-3. Denoting by $u_{i,j}^{(k)}$ the solution obtained after the k-th iteration of the previous two-step procedure, and by $u_{i,j}^{(k+1/2)}$ the Crank-Nicholson solution obtained after step (a), the strategy above ensures that the original variational inequality is satisfied at each iteration, that is,

$$\left(u_{i,j}^{(k)} - u_{i,j}^{(k+1/2)}\right)\left(u_{i,j}^{(k+1/2)} - u_{i,j}^{(k)}\right) \leq 0$$

Chapter 8

SIMULATION EXPERIMENTS

In this section , we report the findings of a number of simulation experiments done for the full-observationcase of the impulse control problem.

We consider an exposure of 100 periods (e.g. 100 days), and, for simplification, we assume that the exposed party faces a default from the other party when the underlying matures at a level lower than its initial level. Otherwise, the exposed party recovers the initial level of the underlying.

For f, the instantaneous cost function of the guarantee, we rely on the Merton (1974) valuation of credit-risk (as Merton's set-up is fully compatible with ours) and thus use the theta of a classical Black and Scholes European put. Of course, the vast literature on credit risk could be a basis of refinement for the target cost function, but we provide here a simple experimentation rather than a definite solution. Other cost functions and choice variables necessary are defined underneath. We first consider the impact of changes in economic environment (volatility, interest rates) on the optimal control policy followed by the credit-risk-exposed party and then analyze the impact of different control cost functions on this optimal policy.

1. CHANGES IN PARAMETERS

We analyze the impact of volatility and interest rate changes when the control cost function is $c(\phi) = a + b\phi$ with $a = b = 0.2$, the maturity of the guarantee is 100 and the Geometric Brownian Motions (GBM) used for the simulation have no drift and starting value $x = 1$. Statistics are derived on 20 GBM realizations for each change of variable.

1.1 IMPACT OF VOLATILITY CHANGES

We set $r = 8\%$ and study the impact of increasing the volatility of the underlying GBM from 0.01 to 0.29. An extract of the results is given in Table I. From option pricing theory, we know that the credit risk exposure increases with the volatility. For the same cost function for controls, jumps become therefore relatively less expensive. We would thus expect controls to either become more frequent or larger when volatility increases. Table 1 shows that with this cost function both happen. Frequency (total number of jumps/total number of periods) increases markedly from 0.012 to 0.057. The size of the jumps increases from -0.010 to 0.740. (Negative jumps can be expected with this type of cost function, as explained underneath). The standard deviation of the size becomes smaller at the same time so that the ratio mean size/standard deviation increases: jumps become larger and positive. The negative skewness shows that controls smaller than average tend to happen more often than jumps larger than average, and this phenomenon gets somewhat stronger with increases in volatility. The increase in kurtosis shows that extreme jumps (very small or very large ones) appear more often for higher volatilities. Neither skewness nor kurtosis appear to be monotonic though. Option pricing theory does not give us much insight in arrival times for the controls. Simulations show that the first two moments of arrival times do not seem to be clearly affected by the increase in volatility. The skewness statistic shows that early controls (compared to the mean) happen more with high volatility than with low volatility. The kurtosis statistic shows a slight increase in extreme values too. In other words, controls do not happen earlier on average when volatility increases, but there are more frequent early controls.

1.2 IMPACT OF INTEREST-RATE CHANGES

Using the same control cost function and the same economic environment, but with a volatility of the GBM of 0.02, we analyze the impact of the increase of interest rates from 0.04 to 0.4 (see Table 2 for results). Because the jumps' costs are discounted to today, we expect an increase in late jumps with an increase in interest rates. Because the cost of the guarantee is decreasing in interest rates, controls become relatively more expensive as interest rates increase and we thus expect less or smaller jumps with an increase in interest rates. Note that the two effects may interact and fool our expectations. Our results show that control frequency is not clearly affected by the increase in interest rates. The size tends to decrease on average, in line with our second expectation (the ratio mean size/standard deviation decreases even more significantly).

Controls tend to become smaller and less significant with an increase in interest rates. The high kurtosis for intermediate values of r shows that extreme jumps occur more often for these values. The arrival time of the controls increases even more markedly, while its standard deviation decreases monotonically. As interest rates increase, jumps occur later and later. The skewness of arrival times decreases and becomes negative. Although controls tend to happen later on average, early outliers also become more frequent.

2. CHANGES IN THE COST FUNCTION

It is clear that f and c are linked and somewhat endogenous: a guarantor may trade off a higher insurance premium (f smaller) for a higher cost of imposing jumps (c larger). A credit-risk-exposed party may trade-off a more lax collateral policy for a more aggressive pricing of credit risk. We study the impact of the shape of c on the optimal policy, which will help us understand the trade-off between f and c that follows. We consider simple cost functions of the shape $c(\phi) = a + b\phi$ (a fixed cost + a proportional cost, for example a cost of capital on the new capital required) and $c(\phi) = a + b\,|\phi|$ (a fixed cost + a transaction cost similar to those observed in trading). Note that the programs can be used with more complex functions. As a first example, consider figures 1, 2 and 3. When there are no fixed costs to control, (e.g., $c(\phi) = 0.2\,|\phi|$) a case similar to proportional costs in a portfolio management setting, the number of jumps is high and their size is small as can be expected. Because there are no fixed costs, very small and frequent controls can occur. As we use discretization, we may actually not obtain through our analysis the smallest jumps involved. Note also that the discretization of the numerical analysis does not show jumps as discontinuities while they actually are discontinuous in a continuous setting. Because the credit-risk-exposed party faces always positive costs at controlling, jumps are always positive. This situation may not be realistic in our case. When there is a positive cost of capital, there may be advantages at decreasing the amount of collateral as the capital can be used for other purposes. Negative jumps would then occur and a cost function of the type $c(\phi) = a + b\phi$ may be preferred. We compare the different simple forms possible and the impact of higher fixed costs (intuitively we should obtain less controls), higher variable costs (intuitively we should obtain smaller controls) and symmetric versus asymmetric cost function.

2.1 INCREASE IN FIXED COSTS

Increasing fixed costs affects the optimal control policy. Consider first the case of no proportional costs, that is where $c(\phi) = a$. In our example, when $a = 0.2$, two jumps occur (figures 5 and 6), while if $a = 0.5$ only one does (Figures 7 and 8). Note that overall control is diminished and $u(y, T/2)$ is smaller in the first case for any y (graphs not shown here; e.g., $u(y = 0.8, t = 50) = 0.2$ in the first case,0.5 in the second; $u(1.0, 50) = 0.13$ in the first caše, 0.22 in the second). High fixed costs of control seem to lead to a less efficient and less frequent control policy. Because there are only fixed costs to control, only positive jumps occur. The credit-risk-exposed-party cannot gain at decreasing the underlying process.

When there are fixed and proportional costs ($c(\phi) = a + b\phi$, with $ab \neq 0$), the controls can become negative. Indeed the gain in cost of capital from having a lower collateral may overcome both the fixed cost of the control and the increase in cost of the guarantee linked to a lower underlying process. This situation occurs in our case with $c(\phi) = 0.1 + 0.2\phi$ (Figures 14 and 15). Increasing the fixed costs to 0.2 and 0.5 lead to a higher overall optimal cost of the collateralized credit risk to the guarantor. For example, we have $u(y = 0.8, t = 50) = 0.2$ for $a = 0.1$; 0.32 for $a = 0.2$ and 0.63 for $a = 0.5$. We also have $u(1.00, 50) = 0.12$ for $a = 0.1$; 0.18 for $a = 0.2$ and 0.28 for $a = 0.5$ (note that Figure 22 gives the graph of u for $c(\phi) = 0.2 + 0.2\phi$ as a function of y). At a low fixed cost, the guarantor could profit from negative jumps. For example, a A-rated company could take out some capital from its AAA-subsidiary while the guarantee of the assets in place in the subsidiary would still be enough to obtain the AAA rating. When fixed costs increase, this opportunity disappears (Figures 16 and 17). In our case, only one jump occurs when $a = 0.2$. Still increasing the fixed costs to 0.5 leads to a slightly changed control (larger jump, see Figures 18 and 19) while the cost of the guarantee becomes larger too. Increasing the fixed costs beyond 0.6 leads to an optimal policy where no controls occur: controls have become too expensive to be worthwhile.

2.2 INCREASE IN VARIABLE COSTS

As usual, increasing costs of control increases the overall cost of the optimal policy. Here, we consider $c(\phi) = 0.2 + 0.2\phi$ and $c(\phi) = 0.2 + 0.5\phi$ (Figures 9 to 12). The numerical analysis gives us that $u(y = 0.8, t = 50) = 0.32$ for $b = 0.2$ and 0.43 for $b = 0.5$. We also have $u(1.0, 50) = 0.18$ for $b = 0.2$ and 0.22 for $b = 0.5$. The optimal policy implemented for a specific Brownian as in the figures shows that when variables costs increase, they overcome the fixed costs. Even though the optimal policy is more expensive, more controls occur and some of these controls become negative. This is empirically interesting and somewhat counterintuitive: Optimally, controls become more frequent even though they are more expensive. The increase in frequency arises from the balance between fixed and variable costs. The size of the controls does not seem significantly affected.

2.3 (A)SYMMETRIC COST FUNCTIONS

We compare the optimal policy when $c(\phi) = 0.1 + 0.2\phi$ versus $c(\phi) = 0.1 + 0.2\,|\phi|$ (Figures 14, 15, 20 and 21). Although optimal policies differ greatly in the two cases, numerical analysis gives us very similar optimal costs of the guarantee at $t = 50$, $u(y,t)$ is overall equal in the two cases and becomes smaller for the first control cost function only when y is very large (y larger than 1.40). It is when y is very large indeed that the credit-risk-exposed party may take advantage of negative controls as these do not affect the cost of the guarantee too greatly in that case. In the application of the optimal policy to a specific Brownian motion described in Figures 13, 14 and 15, the guarantor is indeed able to take advantage of a negative control. It cannot do so with the second cost function where control costs are always positive.

2.4 COMPENSATION BETWEEN *F* AND *C*

Theoretically, a change in premium raised by the credit-risk-exposed party (a change in f) may be compensated by a change in the costs of control (a change in c) so that the choices of f and c are interdependent. We try and test this empirically by studying how a fixed increase in f may be compensated by a change in c. For this we consider $f_1 = f + 0.01$. We first consider u the minimum expected cost of optimal policy as given by numerical analysis with f and $c(\phi) = 0.01 + 0.1\phi$ when the interest rate is zero and the volatility is 0.02 ($T = 100$). We then look for the constant d that minimizes the quadratic difference between u and u1 when u_1 is the minimal cost of the guarantee under optimal policy with instantaneous cost f_1 and control cost $c_1(\phi) = c(\phi) + d$. We find that the quadratic difference becomes minimum (but remains positive) at $d = -0.01$. The programs designed can therefore be used by the credit-risk-exposed party to compensate a change in premiums raised by a change in collaterization control costs.

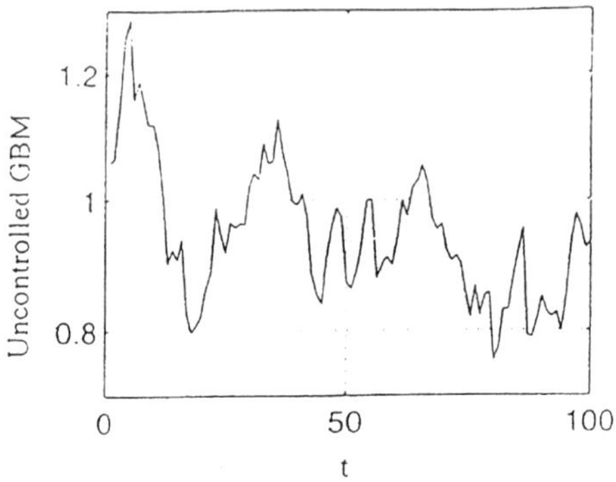

Figure 1. Original uncontrolled process
This graph represents a non-controlled Geometric Brownian Motion with zero drift and 0.05 standard deviation over 100 periods of time (e.g. 4 months for daily returns). The process starts at value 1.

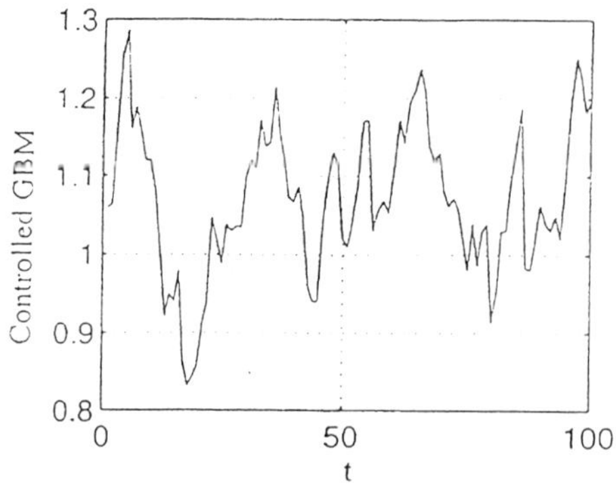

Figure 2. Controlled process: numerical analysis when $c(\phi) = 0.2|\phi|$
This graph represents the same Geometric Brownian Motion with zero drift and 0.05 standard deviation over 100 periods affected by repeated controls, as given by the numerical analysis of the QVI when $c(\phi) = 0.2|\phi|$. Note that the exact controlled process is discontinuous. Because our method relies on discretization this does not appear in the graph.

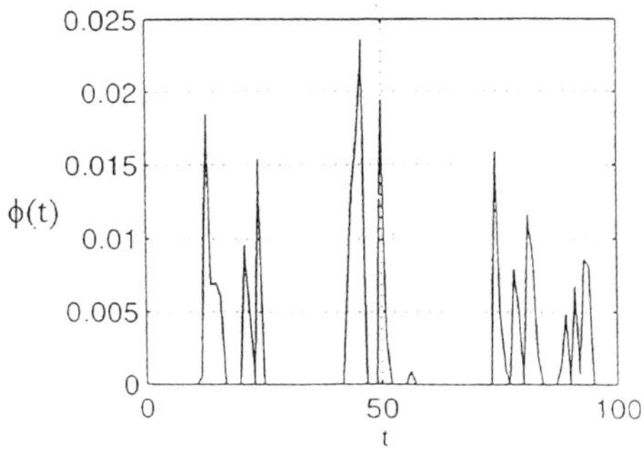

Figure 3. The jumps' process when $c(\phi) = 0.2|\phi|$
This graph represents the process of controls exerted on the Geometric Brownian Motion of figure 1.

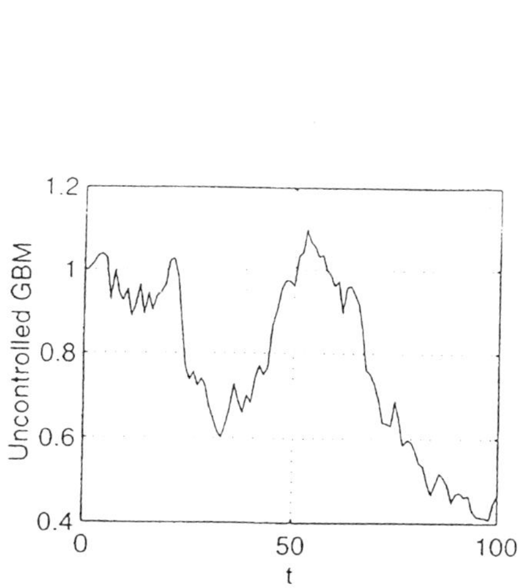

Figure 4. Original uncontrolled process.
Non controlled geometric brownian motion with zero drift, 0.05 standard deviation over 100 periods (e.g. 4 months in daily returns). The process starts at value 1.

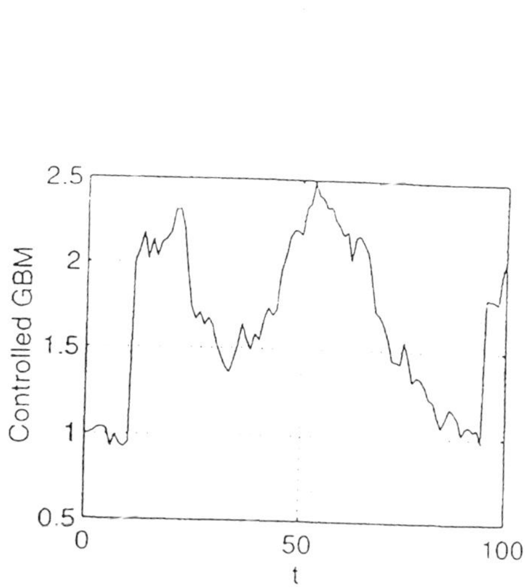

Figure 5. Controlled process.
Numerical analysis applied to the GBM of figure 4 when $c(\phi) = 0.2$
($r = 0.1$, $\sigma = 0.05$, $T = 100$)

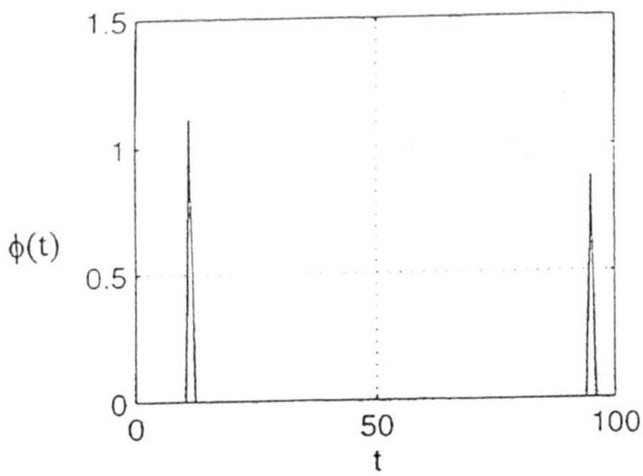

Figure 6. The jump's process
as given by numerical analysis applied to the GBM of figure 4 when c(ϕ) = 0.2 (r = 0.1, σ = 0.05, T = 100).

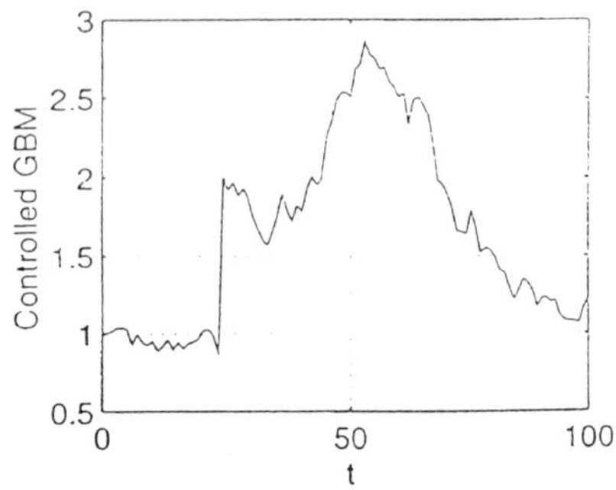

Figure 7. Controlled process.
Numerical analysis applied to the GBM of figure 4 when c(ϕ) = 0.5 (r = 0.1, σ = 0.05, T = 100)

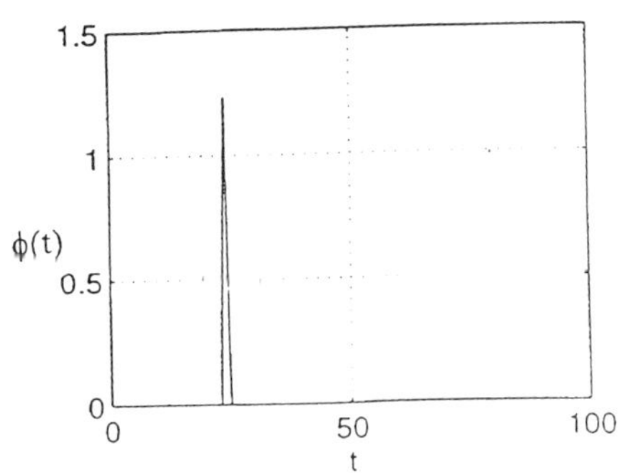

Figure 8. The jump's process given by numerical analysis applied to the GBM of figure 4 when c(ϕ) = 0.5

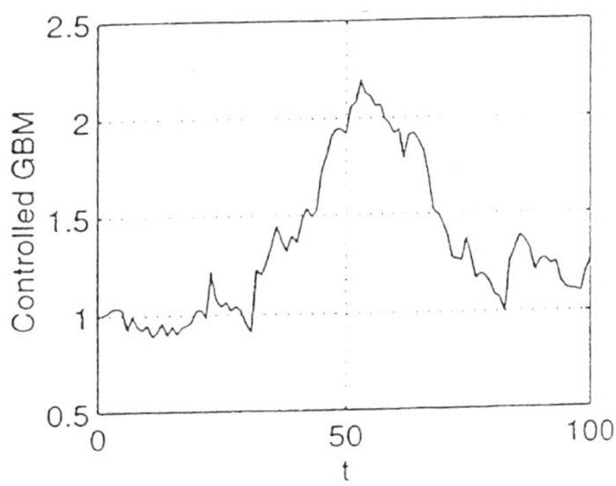

Figure 9. Controlled process: numerical analysis applied to the GBM of figure 4 when $c(\phi) = 0.2 + 0.2\phi$ ($r = 0.1$, $\sigma = 0.05$, $T = 100$)

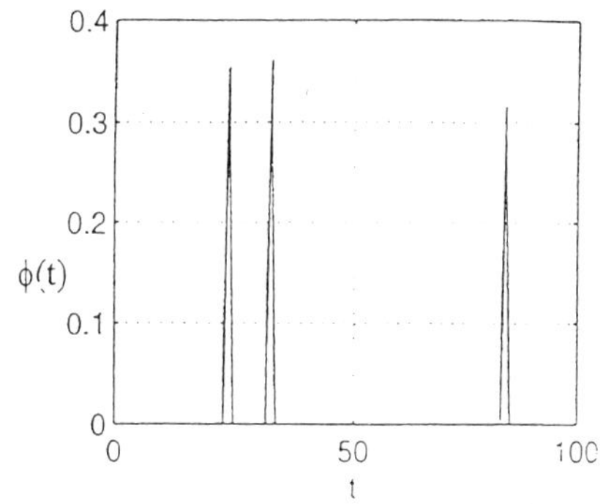

Figure 10. The jump's process as given by numerical analysis applied to the GBM of figure 4 when $c(\phi) = 0.2 + 0.2\phi$.

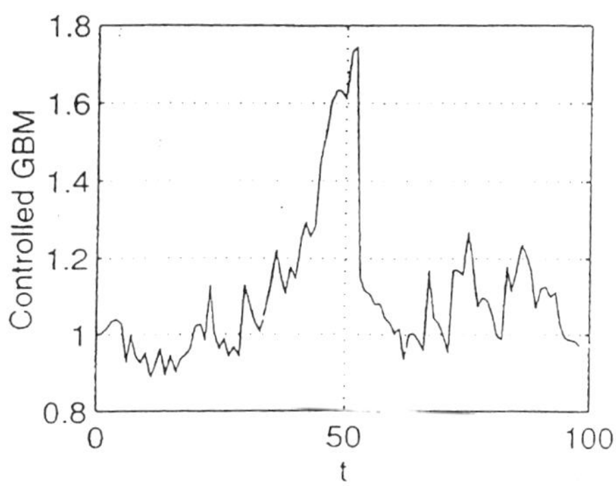

Figure 11. Controlled process: numerical analysis applied to the GBM of figure 4 when $c(\phi) = 0.2 + 0.5\phi$ ($r = 0.1$, $\sigma = 0.05$, $T = 100$).

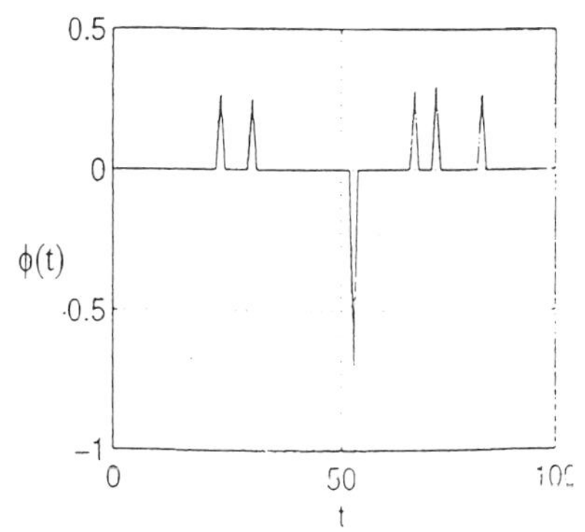

Figure 12. The jump's process as given by numerical analysis applied to the GBM of figure 4 when $c(\phi) = 0.2 + 0.5\phi$.

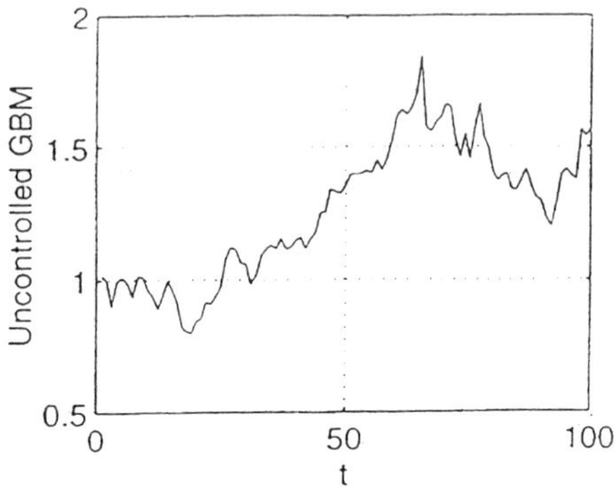

Figure 13. Original uncontrolled process.
Non controlled geometric brownian motion with zero drift, 0.05 standard deviation over 100 periods (e.g. 4 months in daily returns). The process starts at value 1.

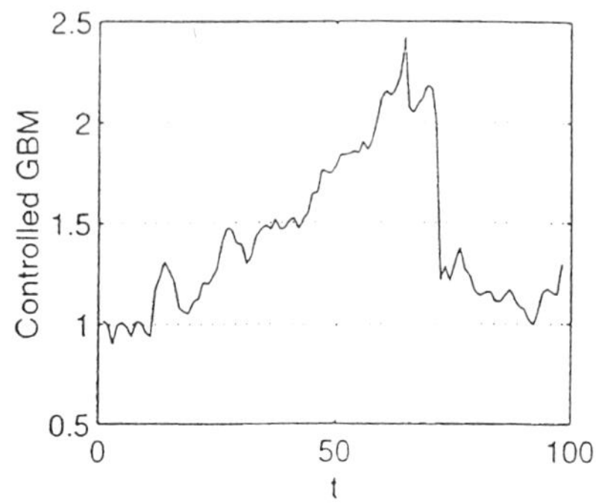

Figure 14. Controlled process: numerical analysis applied to the GBM of figure 13 when $c(\phi) = 0.1 + 0.2\phi$ ($r = 0.1$, $\sigma = 0.05$, $T = 100$)

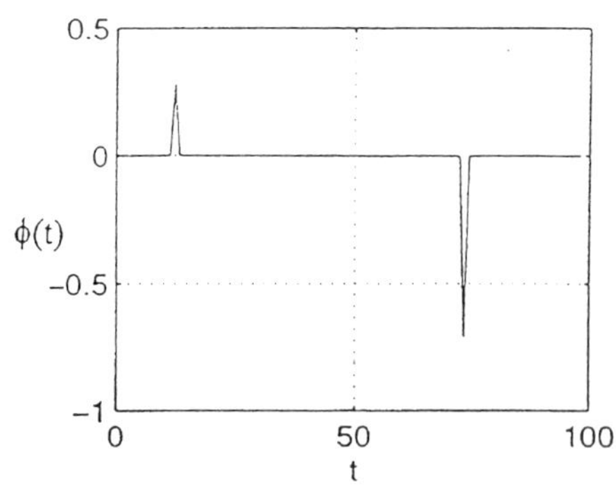

Figure 15. The jump's process as given by numerical analysis applied to the GBM of figure 13 when $c(\phi) = 0.1 + 0.2\phi$

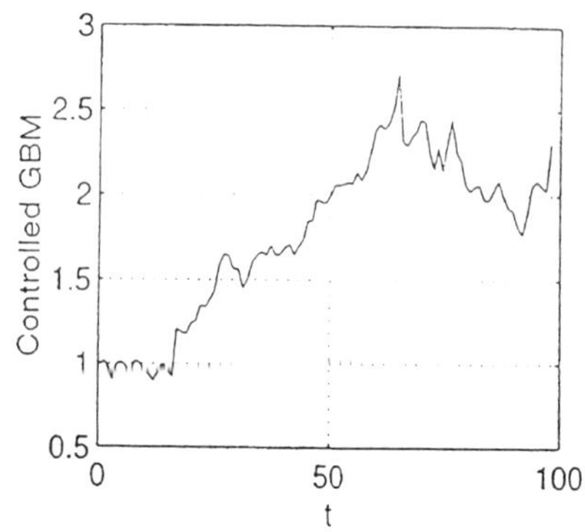

Figure 16. Controlled process: numerical analysis applied to the GBM of figure 13 when $c(\phi) = 0.2 + 0.2\phi$ ($r = 0.1$, $\sigma = 0.05$, $T = 100$)

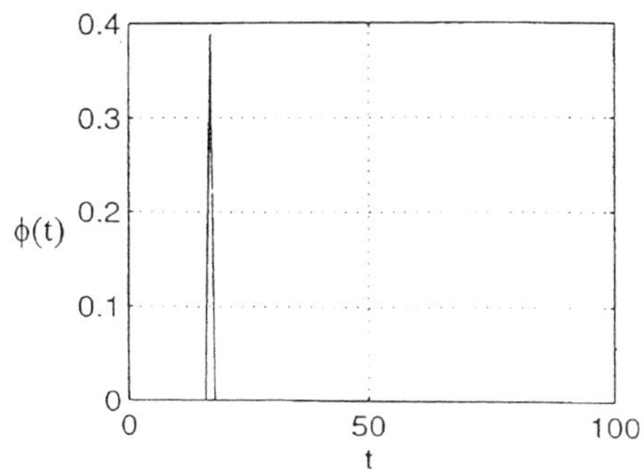

Figure 17. Jump's process given by numerical analysis (GBM of figure 13) when $c(\phi) = 0.2 + 0.2\phi$

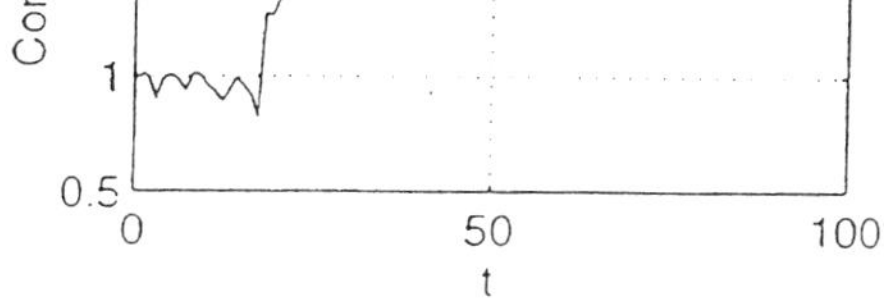

Figure 18. Controlled process: numerical analysis applied to the GBM of figure 13 when $c(\phi) = 0.5 + 0.2\phi$ ($r = 0.1$, $\sigma = 0.05$, $T = 100$)

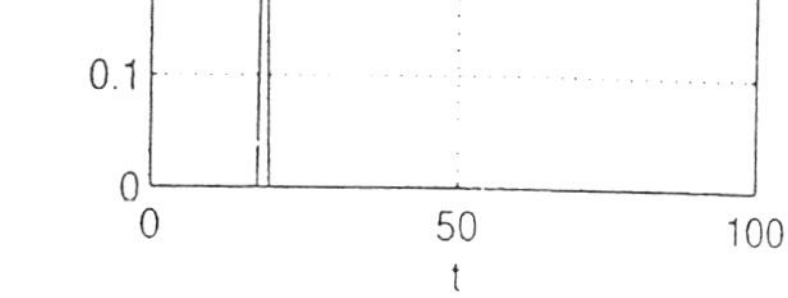

Figure 19. Jump's process as given by numerical analysis applied to the GBM of figure 13 when $c(\phi) = 0.5 + 0.2\Phi$

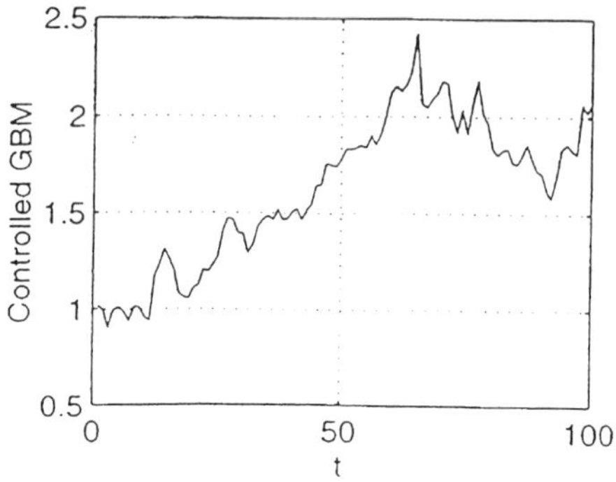

Figure 20. Controlled process: numerical analysis applied to the GBM of figure 13 when $c(\phi) = 0.1 + 0.2|\phi|$ ($r = 0.1$, $\sigma = 0.05$, $T = 100$)

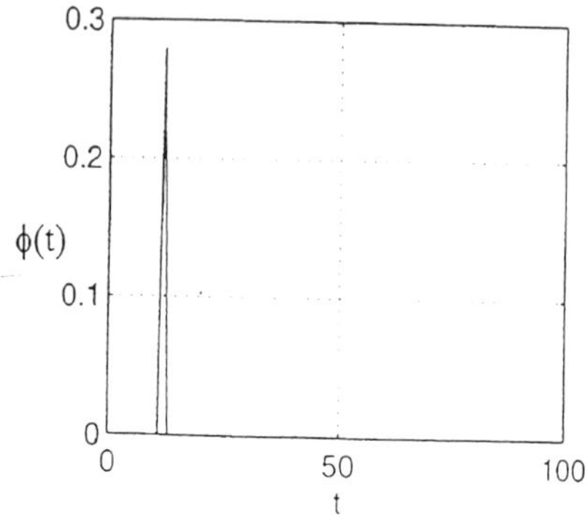

Figure 21. Jump's process as given by numerical analysis applied to the GBM of figure 21 when $c(\phi) = 0.1 + 0.2|\phi|$

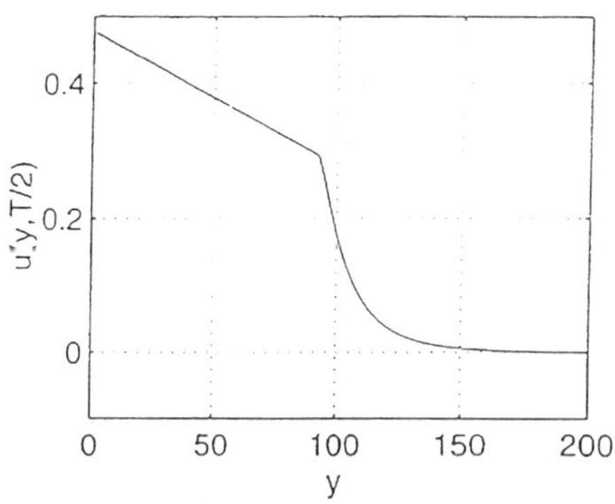

Figure 22. Value of u(y, T/2) as a function of y with $\sigma = 0.05$, $T = 100$, $r = 10\%$, $c(\phi) = 0.2 + 0.2\phi$

Table 1: Impact of Volatility Changes

Impact of changes in volatility of the Geometric Brownian Motion (with 0 drift) on the frequency, size and arrival time of jumps imposed by the optimal control policy. ($T = 100$, 0-drift, $r = 0.08$, $C(\phi) = 0.2 + 0.2\phi$, $y(0) = 1$, statistics taken over 20 different brownian motions with the same characteristics).

		Mean	**Standard Deviation**	**Skewness**	**Kurtosis**
σ = **0.01**	Frequency	0.012			
	Size	-0.010	0.541	-2.266	3.517
	Arrival time	53.79	32.60	0.101	-1.61
σ = **0.03**	Frequency	0.016			
	Size	-0.147	0.838	-1.137	-0.652
	Arrival time	51.59	37.61	0.115	-1.72
σ = **0.05**	Frequency	0.019			
	Size	0.041	0.802	-1.54	0.454
	Arrival time	50.49	35.28	0.043	-1.59
σ = **0.07**	Frequency	0.023			
	Size	0.128	0.784	-1.64	0.781
	Arrival time	52.11	35.08	0.051	-1.56
σ = **0.09**	Frequency	0.023			
	Size	0.343	0.673	-2.42	4.06
	Arrival time	49.26	34.09	0.115	-1.53
σ = **0.11**	Frequency	0.245			
	Size	0.493	0.544	-3.49	10.83
	Arrival time	48.04	32.08	0.149	-1.42
σ = **0.13**	Frequency	0.027			
	Size	0.557	0.597	-3.64	11.81
	Arrival time	46.87	30.41	0.124	-1.32
σ = **0.15**	Frequency	0.029			
	Size	0.622	0.543	-3.776	13.15
	Arrival time	48.07	29.71	0.091	-1.25

σ = **0.17**	Frequency	0.032			
	Size	0.635	0.585	-3.93	14.40
	Arrival time	48.78	29.55	0.028	-1.29
σ = **0.19**	Frequency	0.036			
	Size	0.699	0.376	-4.73	24.99
	Arrival time	49.34	28.76	-0.018	-1.25
σ = **0.21**	Frequency	0.390			
	Size	0.694	0.442	-4.37	20.79
	Arrival time	49.46	28.74	-0.057	-1.26
σ = **0.23**	Frequency	0.045			
	Size	0.644	0.556	-3.64	12.99
	Arrival time	51.67	29.23	-0.077	-1.26
σ = **0.25**	Frequency	0.048			
	Size	0.718	0.425	-4.43	20.43
	Arrival time	49.652	28.28	-0.016	-1.22
σ = **0.27**	Frequency	0.052			
	Size	0.715	0.494	-4.17	17.95
	Arrival time	50.38	28.28	-0.043	-1.21
σ = **0.29**	Frequency	0.057			
	Size	0.740	0.422	-3.77	15.73
	Arrival time	50.92	28.63	-0.045	-1.26

Table 2: Impact of Interest Rate Changes

Impact of changes in interest rates on the frequency, size and arrival time of the jumps imposed by the optimal control policy. (T = 100, 0-drift, $\sigma = 0.02$, $C(\phi) = 0.2 + 0.2\phi$, statistics taken over 20 different brownian motions with the same characteristics)

		Mean	Standard Deviation	Skewness	Kurtosis
r = 0.04	Frequency	0.015			
	Size	0.403	0.789	-0.370	-1.88
	Arrival time	53.65	41.09	0.073	-1.84
r = 0.08	Frequency	0.012			
	Size	-0.112	0.620	-1.444	0.23
	Arrival time	49.13	34.06	0.393	-1.45
r = 0.12	Frequency	0.012			
	Size	0.115	0.243	-4.145	16.00
	Arrival time	51.30	27.64	0.318	-1.22
r = 0.16	Frequency	0.012			
	Size	0.098	0.240	-4.125	15.90
	Arrival time	55.44	26.51	0.201	-1.25
r = 0.20	Frequency	0.010			
	Size	0.088	0.256	-3.802	13.18
	Arrival time	61.40	24.30	-0.157	-0.971

r = 0.24	Frequency	0.010			
	Size	0.082	0.255	-3.775	13.05
	Arrival time	69.50	21.01	-0.530	-0.112
r = 0.28	Frequency	0.011			
	Size	0.077	0.248	-3.885	13.94
	Arrival time	71.90	19.99	-0.448	-0.575
r = 0.32	Frequency	0.011			
	Size	0.024	0.331	-2.64	5.23
	Arrival time	74.32	19.69	-0.529	-0.429
r = 0.36	Frequency	0.010			
	Size	0.068	0.259	-3.67	12.198
	Arrival time	74.53	19.54	-0.629	-0.346
r = 0.40	Frequency	0.012			
	Size	0.020	0.31	-2.804	6.18
	Arrival time	79.71	19.02	-0.919	0.151

Chapter 9

CONCLUSIONS

A general framework for the optimal control of credit risk collaterization was developed that can be applied to deposit insurance or other financial guarantees. The methodology of Quasi Variational Inequalities was adapted to our problem, and exemplifies a possible approach to impulse control problems in finance. Models of optimal auditing and control policies in a variety of situations were derived: full observation with repeated control, full observation with seizure (single control or stopping time), partial observation with repeated control. Numerical analysis of a range of situations helped us characterize the impact of changes in the economic environment (changes in interest rates, changes in volatility) as well as the impact of changes in the control cost functions on the occurrence of controls when the credit-risk-exposed party follows an optimal control policy.

The problem is a complex one and this paper is only but a first try at modeling and analyzing the situation. Some assumptions and simplifications were made that may not be fully realistic but were deemed necessary at this point. Refinements can be added to the model, that would relax these assumptions. Stochastic interest rates, dependence on multiple processes (different risk classes for the credit risk and the collateral risk, or stochastic variance as in Nelson, 1990) or delayed responses to control or audit would be interesting extensions. Even more challenging for future research might be the need to combine continuous-time solutions to control of this type with an equilibrium solution to the asymmetric information and gaming problems that arise in some partial observation situations. In the short term though, first extensions should focus on applications of the model to different empirical situations of interest.

Chapter 10

APPENDIX: PRACTICAL CASES

As an elaboration on the widespread use of guarantees in both private and public finance today, this section presents a survey of situations where guarantees explicitly or implicitly play a major role . It will become obvious from skimming that list that optimal control of guarantees can be an important competitive advantage for some companies, even a necessity when competition is severe, and, in the area of public finance, can save taxpayers significant amounts of money.

There are many uses for guarantees in every financial dimension of today's business world. Some are traditional ways of financing businesses, sometimes through government intervention. Others are at the edge of new financial instruments, exploiting the niches of the introduction of new instruments and alleviating the pricing of their risk. In many cases management of the guarantor will have some dimension of control. In some of these cases, use of our program makes immediate sense as a cheap and efficient alternative to current behavior. In other situations, applications are less obvious and more costly to implement.

1. GUARANTEES IN A PRIVATE SETTING

(i) Financial Security Assurance (FSA) provides insurance to corporate and municipal borrowers in the bond market. Currently, FSA is doing a booming business insuring asset-backed securities. FSA determines what added collateral and surplus cash flow each deal needs to bring the bond to investment grade and arranges for the issuer to place necessary collateral into a special-purpose corporation controlled by FSA. FSA then rents the issuer its triple-A rating, saving the issuer money through lower interest rates and exacting a hefty premium for itself. FSA's guarantee rests on collaterization (or rather over-collaterization). Should the issuer default, FSA can seize the collateral and use it to pay the bondholders. Bond-rating agencies require FSA to have only 25% of the capital needed to take the issue from an enhanced A rating to a AAA rating. For example, in 1988 Bankers Trust led a $200-million debt placement with a triple-A rating for little known S&S Finance International. The debt-laden company (just out of an LBO) had received a guarantee from FSA backed by $332 million in collateral, consisting of shopping centers in prime sites. FSA has the right to acquire the property upon a default and call the bond. Challenges for FSA's business are the collateral's quality, the lack of a resale market for some assets, and the untested nature of the bankruptcy feature of the special-purpose entity. When competitors to FSA truly arise, the need for optimal control will become obvious. Since the contract is idiosyncratic to the situation, it serves as a perfect example of a possible application of our program. Basically, the guarantor can design the control rules (such as situations calling for extension of collateral or seizure of the collateral). Our programs could allow for sensitivity analysis of the different contract designs. Such analysis would enable the guarantor to offer the most efficient contract and meet future competition. (For more on bond insurance, see Spahr, Sunderman, and Amalu (1991). For more on FSA see Barron's, "Financial Alchemy: How FSA Turns Weak Credits into Strong", Jan. 8, 1990, pp.13, 28-29.)

The success of FSA is also an example of a financial intermediary response to a financial market inefficiency(see Merton (1992) and Cossin (1993) on the idea of substitution between markets and institutions). This type of business is likely to encounter some future expansion in the medium term (until a market substitute gets better) as studied by Spahr, Sunderman and Amalu (1991). They show that for the period 1970-1985, the underwriting risk of corporate-bond-guarantee insurance, where the insurer assumes the full obligation of the bond issuer should it default, was less than the normal yield spreads associated with various default ranges. These results suggest that corporate-bond-risk insurance could be offered by a third-party insurer at a lower cost than the prevailing market default-risk premium.

Furthermore, the Tax Reform Act of 1986 reduced tax incentives of municipal bonds for institutional investors. Thus government bonds have been relatively more attractive to individual investors who tend to choose them by their public ratings. Insurance companies have also reduced their premiums on bond insurance, thus creating a rise in demand for bond insurance.

(ii) The use of receivables in corporate finance has changed a lot recently, thanks to securitization, and guarantees have been needed to help this evolution.

Treasury managers seeking property or plant-equipment expansion at non-investment-grade companies have been selling trade receivables, thus creating asset-backed securities. With the aid of credit enhancement such as an insurer's guarantee, companies with junk credit can obtain funds in the same maturity and yield-spread ranges as investment grade companies. (Source: Corporate Cashflow, August 1990, p.33-34.)

Factoring programs put all of the credit risk on receivables in the hands of an asset-management group. Basically, by trusting these assets to another entity, the company gets full guarantee but no more control of its assets (Corporate Cashflow, August 1991, pp. 42). On May 20, 1991, CIT Group/Factoring, a leading factoring firm, recommended that clients halt shipments of new goods to Hess's department stores, thereby implementing basic control. (Wall Street Journal, May 21, 1991, p.8.) It is uncertain how the timing of this control was decided. One can see the costs of either controlling too early (losing a client, etc.) or too late (higher shortfall). Insuring optimal timing thus is dramatically important in such a case.

In 1983, Citicorp launched "conduits"- bank sponsored entities that buy receivables from a number of companies and use them to back ongoing commercial paper or medium-term note programs. Jefferson Smurfit, DynCorp, and Wang Computer have launched issues backed by trade receivables in private placement. This practice obviously allows for financing at lower rates. Banks are required to make capital allocations to guarantee their conduits. (Global Finance, vol.6, no.4, pp.26-27.)

(iii) The use of guarantees in financial markets is also in need of optimal control as described by Szala (1990). She studies three cases of default as focus points of attention on the markets' financial integrity and the purpose of guarantees. In October, 1989, the 135 customer accounts of collapsing options-clearer Fossett Corp. were transferred to a larger clearing firm with the help of an unprecedented \$50-million safety net arrangement. In November 1989, Jordan Sandman Futures Ltd. defaulted on a large margin call, due to the disappearance of a client holding 98% of the short positions. The default prompted the New Zealand Futures Exchange to "invoice back" the price, leaving other members with major losses. Drexel Burnham Lambert's commercial paper default in February 1990, caused its successful London subsidiary to default, but in the United States, 2000 accounts with segregated customer funds of \$400-million were transferred to safe havens before the parent firm's problems expanded. Optimizing control by factoring in the reputation cost for the exchange, as well as other implications of defaults, would lead to optimal minimum-capital requirements for futures commission merchants and possibly new rules of control.

(iv) Other examples of private guarantees in need of optimal control are numerous:

* Loan agreements for aircraft financing are usually collateralized by the aircraft itself. The dynamics of aircraft pricing are well researched by major leasing companies such as GPA. These companies are then in good competitive position to manage and control the guarantees that they offer.

* Commercial paper has become an essential component of many corporate short-term investment portfolios. Today collateralized CP, asset-backed CP and letter of credit CP, with different types for the latter (full and direct pay, irrevocable or stand-by) have increased in importance. The asset-based commercial loan is an important source of business finance. Most of the time the guarantee's asset based consists in the capital stock of the business offered as collateral.

* As described in Savings Institutions, July 1991, p.28-29, mortgage lending as a major portion of a financial's institution business requires an "aggressive" approach to asset and liability management. Prime Bank (Atlanta, Georgia) is the largest residential lender in the Atlanta market. Comparatively better control of its loans has overcome the natural advantage that access to deposits and Federal Home Loan Bank Advances have given savings institutions.

* Financial-performance guarantees are being used widely and successfully throughout the United States in subdivision regulations and other regulatory programs that require developers to construct on- and off-site improvements. Standby letters of credit reduces the exposure to loss of guarantee funds to other creditors. Surety bonds and cash escrows are becoming less dominant.

* Captive finance subsidiaries allow companies to obtain better spreads. The parent guarantees the profitability and the cash flows of the subsidiary. The subsidiary loan agreements include limitations that protect lenders' investment in the subsidiary. (See Ronen, Sondhi, and Birnberg (1989).)

(v) Control of the guarantor:

It should be noted that in some cases, control of the guarantor (rather than of the guaranteed party) could be optimized through a program quite similar to ours. Control of the guarantor (as default of the guarantor jeopardizes the guarantee for the guaranteed party) is an essential and relatively underexplored issue. When in July 1991 New Jersey regulators seized Mutual Benefit Life Insurance Co, the seizure exposed growing problems for investors in certain securities such as industrial revenue and mortgage-backed bonds that carry financial guarantees of insurance companies. Investors were stunned at the time to discover the large amounts of bonds issued in the 1980s with Mutual Benefit guarantees. (Wall Street Journal, Jul. 18, 22, and 23, 1991.) Their relative lack of control certainly did not help their monitoring of the effects of a guarantor's failure. But they still had some control (through buying and selling the securities), and their less-than-optimal behavior proved costly to them.

Similarly, the question of whether holders of guaranteed investment contracts (GIC) will lose everything if the company that issued their GIC fails is primordial to these holders. Some insurers are now offering a version of GICs designed to spread credit risk by placing underlying securities in an account that separates them from the company's general assets. The securities' issuers would have to default to jeopardize the GIC-holder's principal. Still, control of the holders on the underlying securities is limited. (See Financial World, Oct. 29, 1991, pp. 81-82.)

In other cases, control of the guarantor by the guaranteed party is better implemented. For example, risk managers are devoting much of their resources to in-house monitoring of insurers before insolvency problems arise since alternatives after an insolvency are limited. Stock brokerage reports, A.M.Best reports and insurance-broker research are crucial resources for risk managers. The control at hand of risk managers may be options to end the contract earlier (thus the equivalent of seizure) or to decline renewal. In that sense, control can be optimized through a program such as the one we offer.

See also Cook and Spellman (1991) for more on the dependence of guarantee valuation on the financial condition of the guarantor. They specifically study the market's perception of FSLIC's insolvency between 1987 and 1988.

(vi) Other considerations:

It should also be stressed that legal rules and the risk of being unable to implement control should be considered when the program is put into effect. For example, in the Hong Kong case of China & South Seas Bank Ltd versus Tan (1990), a bank loan of HK$30 million was guaranteed by George Tan, who was given a collateral in the form of shares. When repayment was due, the shares were only worth half their sum. There was a long argument between the duty-of-care principle (i.e., the debtor negligently sold the security at an undervalue) and the absence of duty to sell assets at a particular time to lessen the effects of falling prices. The last position won. Clearly optimal control was not allowed to be implemented in that case.

2. GUARANTEES IN A PUBLIC SETTING

The government can provide guarantees for political or foreign policy reasons. The $10 billion in loan guarantees that created a major source of tension between President Bush and Israeli leaders in the fall of 1991, but provided a huge financial boost to Israel, cutting its borrowing costs by up to $600 million over five years, is such an example (*Wall Street Journal*, October 11, 1991, p.6). In these cases, control may not even be an issue. Control will have more importance when economic reasons have led to the use of guarantees. We explore these in the following sections.

(1) Guarantees and Macroeconomic Policy:

(i) Brock (1992) studies how in the 1980s the banking systems of a number of economies collapsed in the face of external shocks. Macroeconomic stabilization efforts were followed by investment booms before the collapses. The investment booms were made possible by government guarantees on foreign loans extended to the economies. Brock presents an analytical model in which a government's delay in the payment of loan guarantees produces an inter-temporal shifting of investment expenditure into the period preceding the termination of guarantees. He examines case studies of Chile and Texas to illustrate the macroeconomic consequences of delays in seizure or closure of insolvent financial institutions. Optimal control by a regulator should therefore incorporate some dimension of the macroeconomic impact in the program (i.e., of the external cost of high correlation of seizure or closure times).

(ii) Guarantees might also be used to solve macroeconomic difficulties. Japan officials, for example, worry about an overall credit crunch for other Asian countries that would slow down the development of the infrastructure of these countries. A possible solution offered would be for institutions such as the Asian Development Bank to guarantee loans from Japanese Banks, which have the funds but lack the balance-sheet capacity (*Far Eastern Economic Review*, vol.154, no.51, December 19, 1991, pp. 73-74). One can imagine, however, how a mismanagement of such guarantees could lead to enormous difficulties. Today's value of these guarantees would probably be affected by this possibility, unless implicit backing of the government was expected. Then the overall cost to the taxpayers could be important if the guarantees are not well controlled over time.

(2) Help to businesses:

Government programs guaranteeing loans or providing loans (and hence guarantees) for businesses are by far too numerous to be comprehensively reviewed here. We just give an idea of their extension and their variety in both the United States and other nations through the following examples. Some of these programs are concerned with helping small or new businesses. Others are promoting exports, in the line of overcoming risks and difficulties of internationalization. The export programs are sometimes in the best tradition of 17th century mercantilism (i.e., a direct mean of promoting the national economy and of providing national companies a financial advantage over foreign competitors). Such programs can also be a form of support to foreign countries. Still other programs target socially desirable industries, like alternative energy.

(i) Promoting small or new businesses:

The Small Business Administration (SBA) guarantees loans of up to $750 000. The SBA 504 program assists exporters, manufacturers, rural firms, minority-owned firms, and others. Since 1980, the program has provided over $3 billion in financing to about 13,000 borrowers. In a typical deal, the SBA guarantees debentures that can provide up to 40% of the cost of the project. For example, SBA guarantees were used in financing Joseph Cerniglia Winery, an apple winery founded by a successful Abraham and Co. broker (D&B Reports, January 1992, pp.10-11).

With such growth, and its relative economic and political importance, perceived as nurturing the industries of the future, optimal control seems essential. Optimal control could help the program stay alive without getting too costly.

(ii) Promoting exports:

These programs, in a multitude of forms, exist in virtually every country of the developed world. OECD export-import banks are essential in providing banks with guarantees for medium-term financing to capital goods exporters.

In the United States, to help businesses face the need for trade financing, the Export-Import Bank (EXIMBANK) provides loans and guarantees in agreement with private lenders. The proportion of each as a source of company financing depends mainly on regulatory concerns as the Credit Reform Act of 1991 has created a bias toward loan financing. (See *Journal of European Business*, July 1991, pp. 43-46.) It is clear, however, that from our analysis, loans and guarantees are basically the same in terms of control. Some well-known examples follow. On September 20, 1991, the United States Export-Import Bank announced guarantees to support $1.3 billion in United States exports to Mexico's oil company, Pemex. (New York Times, September 20, 1991, p.14) On September 21, 1991, the bank signed an agreement to provide guarantees for up to $2 billion in American exports to Kuwait (*Wall Street Journal*, September 23, 1991).

The Overseas Private Investment Corp. (OPIC) complements the EXIMBANK in offering direct and guaranteed-party loans for United States business ventures in over 115 developing countries (*Business America*, March 25, 1991, pp.18-23). State export programs pioneered by Minnesota in 1983 are also now widely used over the United States. From 1985 to 1991, California funded 320 guarantees supporting $381 million in exports (Source: Nation's Business, December 1990, p.26-27). The United States Agency for International Development's Franchising Guarantee Program provides credit guarantees to encourage small- to medium-sized businesses that will be franchisees in developing countries of United States franchise companies. Bi-country organizations also exist that provide guarantees of the same type (e.g. the Hungarian-American Fund). The Multilateral Investment Guarantee Agency (MIGA) provides guarantees against non commercial risks (war or civil disturbances, expropriation, breach of contract, currency transfers) to promote investment in developing countries. (*Finance & Development*, March 1992, pp.48-49.) In this case, control cannot be optimized through a program such as ours: no impulse is possible to change the variable that affects the value of the guarantee (such as a civil war).

The Export Credit Guarantee Department in the United Kingdom, Compagnie Francaise d'Assurance pour le Commerce Exterieur in France and Belgium's Office National Ducroire were set up to encourage export efforts and are inextricably linked with government trade policies and views on foreign affairs. The United Kingdom government privatized the Cardiff Insurance Services Group (ISG), the short-term wing of the Exports Credits Guarantee Department (ECGD). The impact of such privatization and the enforcement of market prices and optimal control has been of great concern to exporters. Because a number of export credit guarantee agencies in Europe operate as companies rather than government departments, it is likely that competition across states will grow fast, creating a need for an optimal form of control that will become even stronger than in isolation.

Such programs can also arise from bilateral agreements. For example, the United States and Japan announced an agreement in May 1991, under which Tokyo helps finance exports of US goods and services. The Japanese provide $1 billion in insurance guarantees for US exports.

(iii) Promoting other businesses:

Governments have been known to target particular businesses, deemed as politically or strategically necessary, and help them through specific guarantees, such as Chrysler in the United States and ICL in the United Kingdom. Targeting is not limited to particular businesses nor to exports or small businesses. Development of alternative energy sources, for example, is helped this way in the United States. Or socially preferable targets are chosen: three Canadian government loan guarantee or insurance programs are available to bankers to assist them in extending debt financing to aboriginal enterprises that demonstrate some potential of viability but are otherwise unable to satisfy normal banking criteria to obtain a commercial loan. All of these programs provide for a loss-sharing ratio of up to 85% to the federal government and at least 15% to the bank.

There are obviously many other instances of targeted loan and guarantee programs. In any of them, optimal control will not only limit adverse selection but will also make the programs financially cheaper and therefore more widely available.

(3) Large projects financing:

Although larger and larger projects have been made without direct public funds, no large project financing today, particularly large infrastructure projects such as Eurotunnel, can secure financing without a complex mixture of multilateral and private money, using intricate overlays of private and government guarantees and insurance (See Beidleman, Fletcher, and Veshosky, 1990). In Bangkok, for example, there is a desperate need for improvement in the infrastructure, as traffic problems have become escalated. But an annual ceiling on public-sector debt commitments drastically limits government guarantees. Optimal control, in order to limit both premiums and overall costs, is then dramatically important.

There has been a trend in many developing nations to place major public investments, especially for infrastructure projects, into the private sector. Build-Own-Operate-Transfer projects (BOOT or BOT), whereby the private sector operates the plant and transfers ownership to the government after a certain concession time, rely heavily on government guarantees to obtain financing.

Governmental agencies that do not usually provide and manage guarantees can be brought in to help projects within the agencies' sphere of influence. In 1990, HUD provided a loan guarantee to Boston City Hospital for a $153-million construction project for a new facility for the hospital (Source: *Modern Healthcare*, November 19, 1990, p.44).

(4) Government guarantees for deposits, mortgages, pensions, etc.:

Finally, governments provide guarantees that lead to a superior Pareto optimum, such as deposit insurance, that, after the thirties, allowed investors and banks to circumvent the risk of runs and the corresponding externalities on the overall economy.

In the United States, the borrowings of government-sponsored enterprises such as the Federal National Mortgage Association (Fannie Mae), the Federal Home Loan Mortgage Corporation (Freddie Mac) and the Student Loan Marketing Association have grown massively from $37 billion in 1970 to more than $1 trillion in 1991 (*The Economist*, May 4, 1991). These borrowings carry federal guarantees. There is a growing worry that officials lack the means to monitor and control the financial soundness of these hybrid federal-private companies (federally chartered and privately owned). Federal backing creates enormous incentives to take risk. For example, the Higher Education Assistance Foundation (HEAF) was founded in 1976 as a private loan-guarantee organization, to guarantee student loans that state agencies were reluctant to back. HEAF stood ready to insure loans to any school no matter what its default record. In 1988, with many defaults in its portfolio, HEAF triggered an automatic reduction in federal reimbursements of the defaults. It was entitled to receive only 80% on its claims sent to the government. The news of HEAF's difficulties have caused its main creditor, the Student Loan Marketing Association (Sallie Mae), problems. The Federal Agricultural Mortgage Corp. (Farmer Mac) provides guarantees of timely payment of principal and interest to holders of securities backed by pools of agricultural, real-estate loans.

The United States Congress established the Pension Benefit Guaranty Corp. (PBGC) as part of its pension-reform legislation in the Employee Retirement Income Security Act of 1974 (ERISA). The PBGC's pension-insurance programs cover 38 million workers in over 112,000 pension plans (see Curtis (1989).) PBGC covers only defined-benefit plans and guarantees the payment of "basic benefits" for single-employer plans. The PBGC is designed to become self-supporting in the long run. On July 25, 1991, PBGC moved to take over two pension plans at bankrupt Pan Am Corp. because they were underfunded (Washington Post, July 25, 1991, p.1). It announced on January 31, 1992, that it wants companies with financially weak pension plans to increase their contributions by $10 billion to $15 billion over 5 years (*Los Angeles Times*, January 31, 1992, p.1). Similarly, state insurance funds cover the insurance industry and become more and more entrenched in exercising needed control. See Bodie (1990), Bodie (1991), and Bodie, Shoven, and Wise (1987).

Control, in the form of seizure if it is the only control option available, have a tremendous economic and political importance in all of these cases. The Federal Home Loan Bank Board has shown strong control of thrift difficulties by using high collaterals, merging insolvent thrifts with stronger institutions and closing unprofitable branches, and reducing the costs of funds through a series of programs to lower risk premiums of troubled thrifts. Federal Home Loan Bank advances have been quite important in providing long-term funding options for savings associations seeking matched-term financing for loans that could not be readily sold in the secondary markets. The net worth of thrifts has been shown to be a significant determinant of demand for advances. Advances are relied upon more heavily by poorly capitalized thrifts (see Smith,1991).

As already mentioned in the literature review, deposit insurance is a major locus of uses of guarantees. Proposals to reduce the subsidy to risk-taking in the current deposit insurance system and to protect the deposit- fund are to require banks to issue subordinated debt, to require bank stockholders to post surety bonds as collaterals, and to infuse equity when required. All of these measures demand optimization of control, as all are forms of control. On May 13, 1992, FDIC proposed to boost the premiums that banks and thrifts pay for deposit insurance by an average of 22%, from $.23 to $.28 for each $100 deposit. Strong institutions will pay less than those thought to have a higher risk of failing (*Wall Street Journal*, May 13, 1992, p.3.). But the rate structure is still quite far from the fair (risk-based) premiums structure advocated by academics. Optimal control is then crucial.

The issue of public guarantee control is not restricted to the United States, although it takes different forms elsewhere. For example, Belgium has been concerned recently about leveling the field between public banks and private ones. Public banks are not required to meet solvency ratios and receive a free-state deposit guarantee but are required to finance commercially unattractive projects. France has been concerned about the implications of the Credit Lyonnais affairs, a state owned bank that used implicit public guarantees to leverage its risk.

Valuation and control of the corresponding public guarantees are doomed to become urgent.

(5) Other example of government guarantees:

Governments use guarantees in numerous other areas, and the need for optimal control, though less publicized, can be strong there too.

As an example, the State Revolving Fund (SRF) was created by Congress in 1989 to help communities that need to finance wastewater treatment projects, with a need evaluated by the EPA to be of $83.5 billion through 2008. SRFs, which has been funded after 1995 only through repayment of the loans, provides either loans directly or guarantees in situations in which this policy would improve credit-market access.

References

Acharya, S. and J. Dreyfus (1989). "Optimal Bank Reorganization Policies and the Pricing of Federal Deposit Insurance", *Journal of Finance*, 44, pp. 1313-1334.

Anderson, R.F and A. Friedman (1977). "A Quality Control Problem and Quasi-Variational Inequalities", *Archive for Rational Mechanics and Analysis*, pp. 205-252.

Anderson, R. W. and S. M. Sundaresan (1996). "The Design and Valuation of Debt Contracts ", *Review of Financial Studies*, 9, pp. 37-68.

Baiocchi, C. (1975). "Studio di un Problema Quazi-Variazionale Connesso a Problemi di Frontiera Libera", *Boll. U.M.I.* 11, 4 (suppl. fasc. 3), pp. 589-613.

Baldwin, C., D. Lessard and S. Mason (1983). "Budgetary Time Bombs: Controlling Government Loan Guarantees", *Canadian Public Policy*, 9, pp. 338-346.

Barbu, V. (1984). *Optimal Control of Variational Inequalities*, Research Notes in Mathematics, Pitman Advanced Publishing Program.

Beaver, W.H., S. Datar and M.A. Wolfson (1990). "The Role of Market Value Account-ing in the Regulation of Insured Depositary Institutions", Working Paper, Stanford Graduate School of Business, Stanford, CA.

Becketti, S. and D.J. Roberts (1990). "Will Increased Regulation of Stock Index Futures Reduce Stock Market Volatility?", *Federal Reserve Bank Economic Review*, Nov/Dec.

Bellman, R., 1957, *Dynamic Programming*, Princeton University Press.

Bensaid, B., J.P. Lesne, H. Pages, and J. Scheinkman (1992). "Derivative Asset Pricing with Transaction Costs", *Mathematical Finance*, 2, 2, pp. 63-86.

Bensoussan, A. and J.L. Lions (1973). "Nouvelle formulation de problèmes de contrôle impulsionnel et applications", *C.R. Acad. Sci. Paris*, 276, pp. 1189-1192.

Bensoussan, A. and J.L. Lions (1982). *Contrôle Impulsionnel et Inéquations Quasi-Variationnelles*, Dunod.

Berger, A.N., K.K. King, and J.M. O'Brien (1991). "The Limitations of Market Value Accounting and a More Realistic ALternative", *Journal of Banking and Finance*, 15, 4-5, pp. 753-783.

Bester, H. (1994). "The Role of Collateral in a Model of Debt Renegotiation.", *Journal of Money, Credit & Banking*, 26 (1), pp. 72-86.

Bodie, Z., J.B. Shoven, and D.A. Wise, ed. (1987). *Issues in Pension Economics*, The University of Chicago Press.

Bodie, Z. (1990). "Pension Funds and Financial Innovation", *Financial Management*, 19, 3, pp. 11-22.

Bodie, Z. (1991). "Shortfall Risk and Pension Fund Asset Management", *Financial Analysts Journal*, 47, 3, pp. 57-61.

Boyle, P.P. and T. Vorst. (1992). "Option Replication in Discrete Time with Transaction Costs". *Journal of Finance,* 47, 1.

Brock, P.L. (1992). "External Shocks and Financial Collapse: Foreign-Loan Guarantees and Intertemporal Substitution of Investment in Texas and Chile", *American Economic Review*, 82, 2, pp. 168-173.

Bulow, J.I. (1981). "Pension Funding and Investment Policy", Mimeo, Stanford University.

Bulow, J.I. (1982). "What Are Corporate Pension Liabilities?", *Quarterly Journal of Economics*, 97, pp. 435-452.

Bulow, J.I. and M.S. Scholes (1982). "Who Owns the Assets in a Defined Pension Plan?", NBER Working Paper No. 924.

Campbell, T.S. and D. Glenn (1984). "Deposit Insurance in a Deregulated Environment", *Journal of Finance*, 39, 3, pp. 775-787.

Chan, Y.-S., S.I. Greenbaum and A.V. Thakor (1992). "Is Fairly Priced Deposit Insurance Possible?" *Journal of Finance*, 47, 1, pp. 227-245.

Ciarlet, P.G. and J.L. Lions, eds (1989). *Handbook of Numerical Analysis*, North-Holland.

Constantidines, G.M. (1986). "Capital Market Equilibrium with Transaction Costs", *Journal of Political Economy*, 94, 4.

Cook, D.O. and L.J. Spellman (1991). "Federal Financial Guarantees and the Occasional Market Pricing of Default Risk: Evidence from Insured Deposits", *Journal of Banking and Finance*, 15, 6, pp. 1113-1130.

Cooper, I. A. and A. S. Mello (1991). "The Default Risk of Swaps ", *Journal of Finance*, 46, 5, pp. 97-620.

Cossin, D. and H. Pirotte (1997). "Swap Credit Risk : an Empirical Investigation on Transactions Data ", *Journal of Banking and Finance*, 21, pp. 1351-1373.

Cossin, D., and H. Pirotte (1998), "How Well Do Classical Credit Risk Models Fit Swap Transaction Data ? ", *European Financial Management Journal*, 4, 1, pp.65-78.

Cossin, D. and H. Pirotte (2000). *Advanced Credit Risk Analysis: Financial Approaches and Mathematical Models to Assess, Price and Manage Credit Risk*, John Wiley & Sons, London, New York, forthcoming Summer 2000.

Cossin, D. and T. Hricko (2000). "Pricing Credit Risk with Risky Collateral", International Association of Financial Engineers Conference, submitted to *Journal of Financial and Quantitative Analysis.*

Crouhy, M. and D. Galai (1991). "A Contingent Claim Analysis of a Regulated Depositary Institution", *Journal of Banking and Finance*, 15, pp. 73-90.

Cummins, J.D. (1988). "Risk-Based Premiums for Insurance Guaranty Funds", *Journal of Finance*, 38, 4, pp. 823-839.

Curtis, S.S. (1989). "Potential for Crisis: Pension Benefit Insurance and the PBGC", *Federation of Insurance and Corporate Counsel Quarterly*, 39, 2, pp.131-140.

Davis, J.L. and W.C. Dale (1991). "Pricing Deposit Insurance: the Role of Regulatory Closure Policy", Mimeo, SEC, Office of Economic Analysis.

Davis, M.H.A and A.R. Norman (1990). "Portfolio Selection with Transaction Costs", *Mathematics of Operation Research*, 15, 4.

Davis, M.H.A., 1993. *Markov Models and Optimization*, Chapman & Hall.

Dixit, A. (1991). "A Simplified Treatment of the Theory of Optimal Regulation of Brownian Motion", *Journal of Economics Dynamics and Control*, 15, pp. 657-673.

Duffie, D. and T. Sun (1990). "Transactions Costs and Portfolio Choice in a Discrete-Continuous Time Setting", *Journal of Economic Dynamics and Control*, 14, pp. 35-51.

Duffie, D. and M. Huang (1996). "Swap Rates and Credit Quality", *Journal of Finance*, 51, 3, pp. 921-949.

Duffie, D. and K.J. Singleton (1997). "An Econometric Model of the Term Structure of Interest-Rate Swap Yields", *Journal of Finance*, 52, 4, pp. 1287-1321.

Dumas, B. (1991). "Super Contact and Related Optimality Conditions", *Journal of Economic Dynamics and Control*, 15, pp. 675-685.

Dumas, B. and E. Luciano (1991). "An Exact Solution to a Dynamic Portfolio Choice Problem Under Transactions Costs", *Journal of Finance*, 46, 2, pp. 577-595.

Eastham, J.F. and K.J. Hastings (1988), "Optimal Impulse Control of Portfolios", *Mathematics of Operations Research*, 13, 4.

Flannery, M.J. and C.M. James (1984). "Market Evidence on the Effective Maturity of Bank Assets and Liabilities", *Journal of Money, Credit, and Banking*, 16, 4, pp. 435-445.

Fleming, W.H. and H.Mete Soner (1993). *Controlled Markov Processes and Viscosity Solutions*, Springer-Verlag.

Flood, R.P. and P.M. Garber (1989). "The Linkage between Speculative Attack and Target Zone Models of Exchange Rates", NBER Working Paper, No. 2918.

Fleming, W.H. and H.M. Soner (1993), *Controlled Markov Processes and Viscosity Solutions*, Springer-Verlag.

Flood, R.P. and P.M. Garber (1989). "The Linkage between Speculative Attack and Target Zone Models of Exchange Rates", NBER Working Paper No. 2918.

Garman, M.B. and J.A. Ohlson (1981). "Valuation of Risky Assets in Arbitrage-Free Economies with Transaction Costs", *Journal of Financial Economics*, 9.

Gennotte, G. and D. Pyle (1991). "Capital Controls and Bank Risk", *Journal of Banking and Finance*, 15, pp. 805-824.

Giammarino, R., E. Schwartz, and J. Zechner (1989). "Market Valuation of Bank Assets and Deposit Insurance in Canada", *Canadian Journal of Economics*, 22, 1, pp. 109-127.

Giberti, D., M. Mentini, and P. Scabellone (1991). "The Valuation of Credit Risk in Swaps: Methodological Issues and Empirical Results", Economics and Finance Series, 11, CRF-IMI Group.

Glowinski, R., J.L. Lions and R. Trémolières (1981). *Numerical Analysis of Variational Inequalitites*, North Holland.

Gorton, G. and A. Santomero (1990). "Market Discipline and Bank Subordi-nated Debt", *Journal of Money, Credit, and Banking*, 22, 1, pp. 119-128.

Greenwald, B.C. and J.C. Stein (1991). "Transactional Risk, Market Crashes, and the Role of Circuit Breakers", *Journal of Business*, 64, 4, pp. 443-462.

Grossman, S.J. and G. Laroque (1990). "Asset Pricing and Optimal Portfolio Choice in the Presence of Illiquid Durable Consumption Goods", *Econometrica*, 58, 1, pp. 25-51.

Hanouzet, B. and J.L. Joly (1979). "Convergence Uniforme des Itérés Définissant la Solution d'Inéquations Quasi-Variationnelles et Application à la Régularité", *Num. Funct. Anal. and Optimiz.*, 1, 4, pp. 399-414.

Harrison, J.M., T.M. Sellke and A.J. Taylor (1983). "Impulse Control of Brownian Motion", *Mathematics of Operations Research*, 8, 3, pp. 454-466.

Hindy, A., C. Huang and S.H. Zhu, S.H. (1997). "Numerical Analysis of a Free-Boundary Singular Control Problem in Financial Economics", *Journal of Economics Dynamics and Control*, 21, pp. 297-327.

Hirtle, B. (1987). "The Growth of the Financial Guarantee Market", *FRBNY Quarterly Review*, pp. 10-28.

Hlavácek,I., J. Haslinger, J. Necas, and J. Lovísek (1988). *Solution of Variational Inequalities in Mechanics*, Springer-Verlag.

Hodges, S.D. and A. Neuberger (1989). "Optimal Replication of Contingent Claims Under Transactions Costs", *The Review of Futures Markets*, 8.

Hull, J. (1989). "Assessing Credit Risk in a Financial Institution's Off-Balance Sheet Commitments." *Journal of Financial and Quantitative Analysis*, 24, 4, pp. 489-500.

Hsieh, S-J., A. H. Chen and K.R. Ferris (1994). "The Valuation of PBGC Insurance Premiums Using an Option Pricing Model", *Journal of Financial and Quantitative Analysis*, 29, 1, pp. 89-99.

Jarrow, R. and S. Turnbull (1995). "Pricing Derivatives on Financial Securities Subject to Credit Risk ", *Journal of Finance*, 50, 1, pp. 53-85.

Jarrow, R., A.D. Lando, and S.M. Turnbull (1997), "A Markov Model for the Term Structure of Credit Risk Spreads", *Review of Financial Studies*, 10, 2, pp. 481-523.

Jones, E.P. and S.P. Mason (1980). "Valuation of Loan Guarantees", *Journal of Banking and Finance*, 4, pp. 89-107.

Jordan, J.V. and G.E. Morgan (1990). "Default Risk in Futures Markets: the Customer-Broker Relationship", *Journal of Finance*, 45, 8, pp. 909-933.

Kalos, M.H. and P.A. Whitlock (1986). *Monte Carlo Methods*, John Wiley & Sons, New-York.

Kane, E.J. (1985). *The Gathering Crisis in Federal Deposit Insurance*, MIT Press, Cambridge, MA.

Kane, E.J. (1986). "Appearance and Reality in Deposit Insurance: the Case for Reform", *Journal of Banking and Finance*, 10, pp. 175-188.

Kane, E.J. (1990). "Principal Agent Problems in S&L Salvage", *Journal of Finance*, 45, 8, pp. 755-765.

Karatzas, I. (1989). "Optimization Problems in the Theory of Continuous Trading", *SIAM Journal of Control and Optimization*, 27, 6, pp.1221-1259.

Keeley, M.C. (1990). "Deposit Insurance, Risk, and Market Power in Banking", *American Economic Review*, pp. 1183-1200.

Kuester, K. and J. O'Brien (1990). "Market-Based Deposit Insurance Premiums: an Evaluation." mimeo, Board of Governors of the Federal Reserve System, Washington, DC.

Laetsch, T.H. (1975). "A Uniqueness Theorem for Elliptic Q.V.I.", *Journal of Functional Analysis*, 18, pp. 286-288.

Lambert, J.D. (1991). *Numerical Methods for Ordinary Differential Systems: the Initial Value Problem*, John Wiley & Sons.

Langetieg, T.C., M.C. Findlay, and L.F.J. da Motta (1982). "Multiperiod Pension Plans and ERISA", *Journal of Financial and Quantitative Analysis*, 17, pp. 603-631.

Leland, H.E. (1985). "Option Pricing and Replication with Transaction Costs", *Journal of Finance*, 40, 5.

Leland, H. and K. Toft (1996), "Optimal Capital Structure, Endogenous Bankruptcy, and the Term Structure of Credit Spreads", *Journal of Finance*, 51, pp. 987-1019.

Longstaff, F. and E. Schwartz (1995), "A Simple Approach to Valuing Risky Fixed and Floating Rate Debt", *Journal of Finance*, 50, 3.

Magill, M. and Constantidines, G. (1976). "Portfolio Selection with Transaction Costs", *Journal of Economic Theory*, 13, pp. 245-263.

Marcus, A. (1987). "Corporate Pension Policy and the Value of PBGC Insurance" in *Issues in Pension Economics*, Bodie, Z., J.B. Shoven, and D.A. Wise, editors.

Marcus, A.J. and I. Shaked (1984). "The Valuation of FDIC Deposit Insurance Using Option-pricing Estimates", *Journal of Money, Credit, and Banking*, 16, 4, pp. 446-460.

Mella-Barral, P. and W. Perraudin (1997). "Strategic Debt Service", *Journal of Finance*, 52, 2, pp.531-556.

Merton, R.C. (1969). "Lifetime Portfolio Selection Under Uncertainty: the Continuous-Time Case", *Review of Economics and Statistics*, 51, pp. 247-257.

Merton, R.C. (1971). "Optimum Consumption and Portfolio Rules in a Continuous-Time Model", *Journal of Economic Theory*, 3, pp. 373-413.

Merton, R.C. (1973). "The Theory of Rational Option Pricing", *Bell Journal of Economics and Management Science*, 4, pp. 141-183.

Merton, R.C. (1976). "Option Pricing When Underlying Stock Returns are Discontinuous", *Journal of Financial Economics*, 3, pp. 125-144.

Merton, R.C. (1977). "An Analytic Derivation of the Cost of Deposit Insurance and Loan Guarantees: an Application of Modern Option Pricing Theory", *Journal of Banking and Finance*, 1, pp. 3-11.

Merton, R.C. (1978). "On the Cost of Deposit Insurance When there are Surveillance Costs", *Journal of Business*, 51, pp. 439-452.

Merton R.C. (1989). "On the Application of the Continuous-Time Theory of Finance to Financial Intermediation and Insurance", *The Geneva Papers on Risk and Insurance*, 14, 52, pp. 225-262.

Merton, R.C. (1990). *Continuous-Time Finance*, Basil Blackwell.

Merton, R.C. (1990). "Capital Market Theory and the Pricing of Financial Securities" in *Handbook of Monetary Economics*, B.M.Friedman and F.H. Hahn, editors.

Merton, R.C. (1990). "The Financial System and Economic Performance", *Journal of Financial Services Research*, 4, pp. 263-300.

Merton, R.C. (1992). "Operation and Regulation in Financial Intermediation: a Functional Perspective", Harvard Business School Working Paper, WP 93-020.

Merton, R. C. (1974). "On the Pricing of Corporate Debt: the Risk Structure of Interest Rates", *Journal of Finance*, 29, pp. 449-470.

Merton, R.C. and Z. Bodie (1992). "On the Management of Financial Guarantees", *Financial Management*, 22.

Osterberg, W.P. and J.B. Thomson (1991). "The Effect of Subordinated Debt and Surety Bonds on the Cost of Capital for Banks and the Value of Federal Deposit Insurance", *Journal of Banking and Finance*, 15, pp. 939-953.

Pecchenino, R.A. (1992). "Risk-based Deposit Insurance: an Incentive Compatible Plan", *Journal of Money, Credit and Banking*, 24, 4, pp. 499-510.

Pennacchi, G. (1987a). "A Reexamination of the Over- (or Under-) Pricing of Deposit Insurance", *Journal of Money, Credit and Banking*, 19, pp. 340-360.

Pennacchi, G. (1987b). "Alternative Forms of Deposit Insurance: Pricing and Bank Incentive Issues." *Journal of Banking and Finance*, vol. 11, pp. 291-312.

Perold, A. (1988). "The Implementation Shortfall: Paper versus Reality", *Financial Analysts Journal*.

Pratt, J.W. and R.J. Zeckhauser eds. (1985). *Principal and Agents: the Structure of Business*, Harvard Business School Press.

Reinhardt, U. (1973). "Break-Event Analysis for Lockheed's TriStar: an Apllication of Financial Theory", *Journal of Finance*, 28, pp. 821-838.

Index

Advances in Computational Management Science

---

1. J.-M. Aurifeille and C. Deissenberg (eds.): *Bio-Mimetic Approaches in Management Science.* 1998 ISBN 0-7923-4993-8
2. A.-P.N. Refenes, A.N. Burgess amd J.E. Moody (eds.): Decision Technologies for ComputationalFinance. Proceedings of the Fifth International Conference Computational Finance. 1998 ISBN 0-7923-8308-7 (HB); ISBN 0-7923-8309-5 (PB)
3. D. Cossin and F.M. Aparicio (authors): *Optimal Control of Credit Risk.* 2001 ISBN 0-7923-7938-1